Agroecology Now!

Colin Ray Anderson • Janneke Bruil
M. Jahi Chappell • Csilla Kiss
Michel Patrick Pimbert

Agroecology Now!

Transformations Towards More Just
and Sustainable Food Systems

Colin Ray Anderson
Centre for Agroecology, Water and
Resilience
Coventry University
Wolston, UK

M. Jahi Chappell
Centre for Agroecology, Water and
Resilience
Coventry University
Coventry, UK

Michel Patrick Pimbert
Centre for Agroecology, Water and
Resilience
Coventry University
Coventry, UK

Janneke Bruil
Cultivate! Collective
Bennekom, The Netherlands

Csilla Kiss
Centre for Agroecology, Water and
Resilience
Coventry University
Coventry, UK

ISBN 978-3-030-61314-3   ISBN 978-3-030-61315-0   (eBook)
https://doi.org/10.1007/978-3-030-61315-0

Cover pattern © Melisa Hasan

This Palgrave Macmillan imprint is published by the registered company Springer Nature Switzerland AG.
The registered company address is: Gewerbestrasse 11, 6330 Cham, Switzerland

ACKNOWLEDGEMENTS

We are grateful to Jessica Milgroom for inputs, editing and ideas that helped to shape this book in the late phases of its development and to Annelie Bernhart, Chris Maughan and Diana Quiroz for their helpful contributions and comments on particular sections. We greatly appreciate the support and inputs of Beate Scherf, Maryam Rahmanian, Remi Cluset, Emma Siliprandi, Soren Moller and Caterina Batello at the Food and Agriculture Organization of the United Nations (FAO). We also thank colleagues who prepared the substantial and detailed case studies on aspects of agroecology transitions that we drew on for boxes and sections: Mindi Schneider, Xu Ye, Chantal Jacovetti, Julia Wright, Iain MacKinnon, Graciela Romero Vasquez, Pedro Lopez Merino, Patrick Mulvaney, Nils McCune. And further thanks are due to Million Belay and Jan Douwe van der Ploeg for reviewing and commenting on earlier drafts.

We would like to express our appreciation to the participants of an international workshop organized to review this text and discuss the role of agroecology in transitions to sustainable food systems: Mindi Schneider, Claire Lamine, Paulo Petersen, Marta Rivera Ferre, Paola Migliorini and Andrea Ferrante. Thank you to the 14 FAO staff who attended the international workshop and offered their feedback on this text. This workshop, and the interactions around the topic more generally, was a rich exchange largely because it drew together the perspectives on agroecology transitions of the team at Coventry University's Centre for Agroecology,

To access supplementary learning materials, videos, podcasts and more visit: https://www.agroecologynow.com/transformation.

Water and Resilience (CAWR), who had been intensely studying agroecology transitions with those of external reviewers and FAO staff members. The debates have been energizing, often surprising and a great learning experience for all, and have greatly improved both our thinking and this book.

Finally, we would like to thank the diligent professionals who helped to bring this book to fruition. Our gratitude goes out to Barbara Kiser for her diligent and thoughtful review and close edit of our initial manuscript and to the team at Palgrave Macmillan—especially G. Nirmal Kumar, Joanna O'Niell and Rachael Ballard.

Coventry, 2020

Colin Ray Anderson
Janneke Bruil
Michael Jahi Chappell
Csilla Kiss
Michel Pimbert

CONTENTS

LIST OF FIGURES

LIST OF BOXES

CHAPTER 1

Introduction

Abstract In this introductory chapter, we introduce agroecology as an urgent alternative paradigm for food and farming in a time of growing ecological, economic and social crises. We briefly outline the role of food systems in these intersecting crises and introduce how agroecology is much more than a 'technical fix' that calls to tweak the existing system. It is rather a framework for transformation that can be adopted in pursuit of a more just and sustainable food system. The chapter describes the origin of the book and provides a roadmap to help the reader navigate the flow of the manuscript.

Keywords Agroecology • Transformation • Crisis • Social movements

AGROECOLOGY: AN IDEA FOR URGENT TIMES

In her recent analysis of the COVID-19 pandemic, Canadian journalist Naomi Klein ironically riffs off a famous phrase on crises by free-market economist Milton Friedman. When catastrophe hits, he noted, "the actions that are taken depend on the ideas that are lying around" (Klein 2020). In the context of current and imminent crises—from climate change and biodiversity loss to hunger, poverty and disease—it is clear that catastrophe is not only on our doorstep but has arrived for many peoples around the world. It is also clear that agroecology is not just an idea that is 'lying around' but one that has been teed up by visionary food producers, social movements and researchers. The time for agroecology is now.

Over the past five years, the theory and practice of agroecology have crystalized as an alternative paradigm and vision for food systems. Agroecology is an approach to agriculture and food systems that mimics nature, stresses the importance of local knowledge and participatory processes and prioritizes the agency and voice of food producers over corporations and other elite actors. As a traditional practice, its history stretches back millennia, whereas a more contemporary agroecology has been developed and articulated in scientific and social movement circles over the last century. Most recently, agroecology—practised by hundreds of millions of farmers around the globe—has become increasingly viewed as viable, necessary and politically possible as the limitations and destructiveness of 'business as usual' in agriculture have been laid bare.

But as a system, agroecology has powerful competition in the corporate actors who peddle high-tech, profit-centred 'solutions' that preserve an unjust and unsustainable food system and agroeclogy remains marginal, its potential effectively sabotaged by the political interests that continue to embolden the high-input industrial model. The battle for the future of food and farming is intensifying with the growing sense of urgency over our intermeshed ecological and social crises.

There is now much evidence to show that our socio-economic systems are catastrophically undermining the function of natural systems. The Intergovernmental Panel on Climate Change (IPCC) (2019) notes that between 2007 and 2016 some 23% of total anthropogenic greenhouse gas emissions came from unsustainable practices in agriculture, forestry and other land-use activities. Other major reports have drawn attention to convergent crises such as accelerating extinction rates (IPBES 2019), looming water shortages for five billion people (World Water Assessment Programme (WWAP) 2019), UNESCO rising world hunger (FAO 2019), dangerous degradation and pollution of land and soil, mounting resource depletion and a rise in levels of air pollution resulting in disease and health-related death (Health Effects Institute 2018). And, most recently, the COVID-19 crisis has revealed the vulnerability that arises from a just-in-time, centralized industrial food system (Wallace et al. 2020).

In fact, the pandemic has revealed how industrial agriculture contributes to the rise and spread of deadly pathogens by pushing agriculture and extraction further into the forest and by creating densely crowded genetically homogenous domestic livestock populations that are breeding grounds for the emergence of zoonotic viruses (Wallace 2016; Wallace et al. 2020). 'Industrial food', as a system, both spawns large-scale ecological, social and economic problems and reduces the capacity or resiliency of farmers and communities to cope with change. Major shifts are needed, not tweaks to the failing system we have.

Despite significant underfunding and lack of research (see Chap. 5), evidence on the multifunctional benefits of agroecology are growing (summarized in Chap. 2). In contrast, agroecology represents a system that works with nature instead of against it and offers an approach to food production that boosts biodiversity, creates ecological resilience, improves soils, cools the planet and reduces energy and resource use. It has been shown to be highly productive, to provide highly diverse dietary offerings and to support the process of community building and women's empowerment (Fig. 1.1).

Fig. 1.1 *The film Agroecology:* Voices from *Social Movements* exemplifies the book's primary theme: the struggle to advance agroecology as an alternative to the dominant food regime.View video here: https://www.agroecologynow.com/video/ag/ (Photo credit: Authors)

The agroecology that we embrace in this book emerges not only as an alternative to the oft-critiqued industrial and corporate food system, however. It must also be part of the effort to counter racial capitalism, patriarchy and other forms of structural violence and oppression. Although anti-racism, indigenous cosmovision, decolonization and feminism are often found only in the radical margins of the agroecology canon, it is in these traditions that the transformative potential of agroecology can be deepened. Movements from Black Lives Matter to the World March of Women offer potential lessons and allies for agroecology. So do on-the-ground experiments with equity and radical democracy, such as those taking place in the autonomous region of Rojava in Syria, and the work of action researchers exploring decoloniality, feminist political ecology, queer ecology, critical physical geography and beyond.

In this context, a transformative agroecology can be imagined as one manifestation of a global struggle for emancipation—achievable through solidarities, ally-ship and strategic action. Thus, while food systems are this book's focus, we make connections throughout to the intersection with wider struggles against oppressions and call for the field and practitioners of agroecology to integrate further with these wider movements for change.

A deeply politicized and collectivized practice of building agroecology from the bottom up is, we argue, the essential *basis* for transformation in food systems. We believe that this will happen only when the dominant regime is itself transformed to enable agroecology as *an objective of transformation*. The dialectical process is central to the aim. We take an agency-centric approach (see, e.g., the discussion of agency in HLPE (2019), working alongside our allies from many walks of life, in facing this challenge to the hierarchies and assumptions of the dominant regime. Our approach identifies the need for substantial shifts in governance and power. If agroecology is indeed a good idea that is lying around, it is time to map out how we can seize the moment for the transition to a more just and sustainable food system, and society.

THE ORIGINS AND PURPOSE OF THE BOOK

This book is the result of a research collaboration that started in 2018 with a literature review and case-study development for the Food and Agriculture Organization of the United Nations (FAO). Our eclectic group of activist scholars set out to understand how to amplify agroecology while moving towards just, sustainable food systems. We analysed academic and non-academic literature to better understand the dynamics involved in agroecology transitions, the opportunities and obstacles, and the role of governance and power. We also invited a small number of people with practical and/or academic experience in agroecology to submit new case studies as a way of deepening our understanding of particular aspects of the field.

At the end of 2018, an initial version of this research was presented at a multiday workshop in Rome involving academics, social-movement leaders and FAO staff from around the world. We continue to be grateful for their suggestions for improvements. Since then, FAO has used our report internally as part of its global policy process to support the scaling up of agroecology.

In 2019, we published snippets of our findings, for example in an article in the journal *Sustainability* and as a 'backgrounder' (Anderson et al. 2019). Following that, we were told by friends and colleagues that a publicly accessible version of the full research paper would be very timely. They encouraged us to use the opportunity to publicize the idea of a transformative approach to agroecology more widely. We are, after all, at a crucial moment in this effort, as agroecology gains traction not only with FAO but also with national governments, social movements and other actors—with the associated risks and opportunities. The idea of an open access publication emerged. Palgrave Macmillan agreed to publish an updated version of our work: the result is this book.

In it, we seek to provide insights into approaches to agroecology, based on core principles adapted to place and context rather than proscriptive rules. We articulate agroecology as an ongoing *process* of food-system transformation, supported by a set of underlying *values* based on ecological principles and social justice, and honouring the *agency* of food producers and the important role of social movements in transformational change. Thus, while our aim is to understand and support large-scale transformational change, our approach is to focus on the tangible changes that are possible when working from the bottom up in communities and social movements. This requires a simultaneous process of strengthening and building agroecology as a radical alternative while also deconstructing the dominant corporate food regime that lock in unsustainable and unjust food systems.

Ultimately, this book aims to serve, directly or indirectly, agroecologists—particularly organizations and networks of agricultural producers, and especially women. Much of the thinking that went into it has been inspired by what we have learned from them. We hope the combination of a theoretical and analytical framework with more empirical analyses (including case studies) will offer intellectual and practical inspiration to academics and students keen to understand how territorial efforts may be connected to system-wide transformations.

As we have noted, our findings will also speak to people in other political movements—from climate and environmental justice to anti-racism, de-growth and feminism. We believe that the insights are relevant too to policy-makers, journalists and other advocates of healthier, more sustainable and accessible food and agriculture systems.

A ROADMAP TO THE BOOK

In Part I, we elaborate on the history, meaning and multiple ecological, economic and social benefits of agroecology. We then introduce the notion of a transformative agroecology rooted in the tradition of political ecology adopted in this book. To better conceptualize the process of transformation, we use the multi-level perspective—an influential framework for analysing sustainability transitions across space and time (Geels 2011; Geels and Kemp 2007). With this approach, we show how agroecology—which emphasizes the agency of people—sits within a dominant regime that operates through deep 'landscape' level processes of capitalism, racism, patriarchy and colonialism. It is in the interface and conflict between these two paradigms that transformation—spurred by collective action, shifts in governance and building of countervailing power—can occur.

In Part II, we introduce the idea of 'domains of transformation', which we flesh out as discrete conceptual areas within which the dominant regime poses barriers to the development of agroecology. On the other hand, it is also within each of these domains that proponents of agroecology are taking collective to advance the transformative project at the heart of agroecology. Thus, the domains represent discrete but deeply interconnected areas where the regime and agroecology collide and where further interventions are required to enable agroecology transformations. Synthesizing the literature and bringing in case studies and vignettes from our research and networks, we present six such domains: rights and access to nature (Chap. 4); knowledge and culture (Chap. 5); systems of economic exchange (Chap. 6); networks (Chap. 7); equity (Chap. 8) and discourse (Chap. 9). However, as will be demonstrated, efforts in one domain alone are insufficient and it is a holistic and integrated approach across all of these domains where the greatest potential for agroecology transformations manifests.

Finally, in Part III, we drill down on issues of governance, power and control across all six domains to find the fundamental drivers of transformation through agroecology. We have identified six distinct ways in which different governance interventions (such as new state policies, the building of new 'nested markets', and the actions of civil society networks) affect the dynamics between the dominant food system and emergent agroecological alternatives. When top-down technocratic approaches in governance shift towards bottom-up distributed ones, agroecology is enabled in all the domains, and ultimately, as the changes in each domain overlap, they will synergize towards a system-wide shift.

References

Anderson, C. R., Bruil, J., Chappell, M. J., Kiss, C., & Pimbert, M. P. (2019). From Transition to Domains of Transformation: Getting to Sustainable and Just Food Systems through Agroecology. *Sustainability, 11*(19).

FAO. (2019). *2019—The State of Food Security and Nutrition in the World (SOFI): Safeguarding against Economic Slowdowns and Downturns*. Rome: FAO.

Geels, F. W. (2011). The Multi-Level Perspective on Sustainability Transitions: Responses to Seven Criticisms. *Environmental Innovation and Societal Transitions, 1*(1), 24–40.

Geels, F. W., & Kemp, R. (2007). Dynamics in Socio-Technical Systems: Typology of Change Processes and Contrasting Case Studies. *Technology in Society, 29*(4), 441–455.

Health Effects Institute. (2018). *State of Global Air 2018*. Special Report. Boston, MA: Health Effects Institute.

HLPE (2019). *Agroecological and other innovative approaches for sustainable agriculture and food systems that enhance food security and nutrition*. Rome: High Level Panel of Experts on Food Security and Nutrition of the Committee on World Food Security.

IPBES. (2019). *Global Assessment Report on Biodiversity and Ecosystem Services*.

IPCC. (2019). *Climate Change and Land: An IPCC Special Report on Climate Change, Desertification, Land Degradation, Sustainable Land Management*.

Klein, N. (2020). Coronavirus Capitalism—And How to Beat It. In The Intercept (Ed.) https://theintercept.com/2020/03/16/coronavirus-capitalism/.

Wallace, R. (2016). *Big Farms Make Big Flu: Dispatches on Influenza, Agribusiness, and the Nature of Science*. New York, NY: Monthly Review Press.

Wallace, R., Liebman, A., Fernando Chaves, L., & Wallace, R. (2020). COVID-19 and Circuits of Capital. *Monthly Review, 72*(1).

WWAP (UNESCO World Water Assessment Programme). (2019). The United Nations World Water Development Report 2019: Leaving No One Behind. In UNESCO (Ed.). Paris.

PART I

Agroecology and Sustainability
Transformations

CHAPTER 2

Origins, Benefits and the Political Basis of Agroecology

Abstract In this chapter, we introduce the origins and history of agroecology, outlining its emergence as a science and its longstanding history as a traditional practice throughout the world. We provide a brief review of the evidence of the benefits of agroecology in relation to productivity, livelihoods, biodiversity, nutrition, climate change and enhancing social relations. We then outline our approach to agroecology which is rooted in the tradition of political ecology that posits power and governance have always been the decisive factors in shaping agricultural and other 'human' systems.

Keywords Multifunctional benefits • Ecology of food systems • Food sovereignty • Power • Political ecology • Governance

HISTORY OF AGROECOLOGY

In this book, we focus on the notion of agroecology as a substantial departure from the solutions to today's crises being proposed by mainstream actors. These solutions include technology- and corporate-led societal transformation based on large-scale interventions (e.g. geoengineering to cool the planet), new technologies (e.g. artificial intelligence and robotics) and market-led solutions to drive sustainability transitions (World Economic Forum 2018). At the same time, there is growing support for civil-society led processes of self-organization like agroecology (IPES-Food 2016; Nyeleni 2015). These bottom-up transformations are already

© The Author(s) 2021
C. R. Anderson et al., *Agroecology Now!*,
https://doi.org/10.1007/978-3-030-61315-0_2

happening around the world, marking a challenge to the power of actors empowered within the dominant global food system. In this section, we will take a closer look at the evolution of the idea and practice of agroecology.

Within science, agroecology has been seen as an important regenerative form of agriculture and food systems for almost a century, with practices aimed at mimicking or harnessing complex ecological processes (Box 2.1). Miguel Altieri's (2018) definition of agroecology as the application of ecological concepts and principles to the design and management of sustainable agroecosystems has been a key reference point. In the late 1990s, the framing of agroecology within English-language academic writing was broadened, moving beyond the farm to include food production, distribution, consumption and waste management. This led to a new and more comprehensive definition of the study of agroecology as "the ecology of food systems" (Francis et al. 2003).

While this early scientific work on agroecology was fundamental to articulating its ecological dimensions, it did not engage with political ones, which have long been advanced by social movements and farmers' organizations. Importantly, the scientific literature did not adequately acknowledge the deep foundations and precursors of agroecology in traditional and contemporary practices of indigenous peoples and peasant farmers (Hernández Xolocotzi 1977). Nor did it mention political and ecological critiques of the rise of industrialized agriculture by nineteenth-century luminaries such as Peter Kropotkin, Justus von Liebig and Karl Marx (Foster 1999; Kropotkin 2015) or the political, ecological and technical critiques and alternatives by original thinkers including Albert Howard, Eve Balfour, J. I. Rodale and George Washington Carver (Doré and Bellon 2019; White 2018).

Box 2.1   The Production Principles of Agroecology

According to Altieri (2018), agroecology can be understood as the application of ecological concepts and principles to the design and management of sustainable agroecosystems. The following agroecological production principles (drawn from a variety of sources) can work in synergy and provide the basis for the design of sustainable farming systems:

(*continued*)

Box 2.1 (continued)

- Adapting to the local environment
- Building healthy soils rich in organic matter
- Conserving soil and water
- Diversifying species, crop varieties and livestock breeds in the agroecosystem over time and space from a landscape perspective

Fig. 2.1 Indigenous Lepcha farmers in Sikkim saving traditional seeds adapted to place and deeply tied to cultural practices (*Photo credit*: David Meek)

(*continued*)

Box 2.1 (continued)

- Enhancing biological interactions and productivity throughout the system rather than focusing on individual species and single genetic varieties
- Minimizing the use of external resources and inputs (e.g. for nutrients and pest management)

Popular agroecological practices around the world in which these principles are applied include intercropping, agroforestry, no tillage and mulching. In different contexts, specific practices have been developed, such as the 'push and pull' (an approach that uses 'push' plants to repel and trap plants to 'pull's damaging insects) technique for natural pest control used on sorghum and corn in Ethiopia, the ancient Mexican practice of *milpa* (growing squash, corn and beans together), the system of rice intensification much used in Asia and the Sahelian practices of farmer-managed natural regeneration of tree shrubs and zaï holes—digging pits that retain water and nutrients. However, it is now widely recognized that agroecology entails more than such technical aspects and also has strong socio-political dimensions.

Agroecology is sometimes assumed to be an end goal. In actuality, it is—as we have shown—a process of continuous transition based on core principles (Altieri 2018; HLPE 2019), values and politics (Nyeleni 2015) or specific cultural, ecological or social elements (FAO 2018). Such organizing principles have been depicted in lists and infographics and vary in orientation, politics and presentation (see the Agroecology Compass—www.agroecologycompass.net—which aggregates many of these). Like the proliferating definitions of agroecology, only some of these models reflect a deeply transformative perspective.

The Food and Agriculture Organization of the United Nations (FAO) sees agroecology as having ten primary 'elements', from diversity and resilience to human and social values, and focuses on interdependencies between them (Fig. 2.2) (FAO 2018). This breakdown is impressively nuanced in social and political terms, for a mainstream

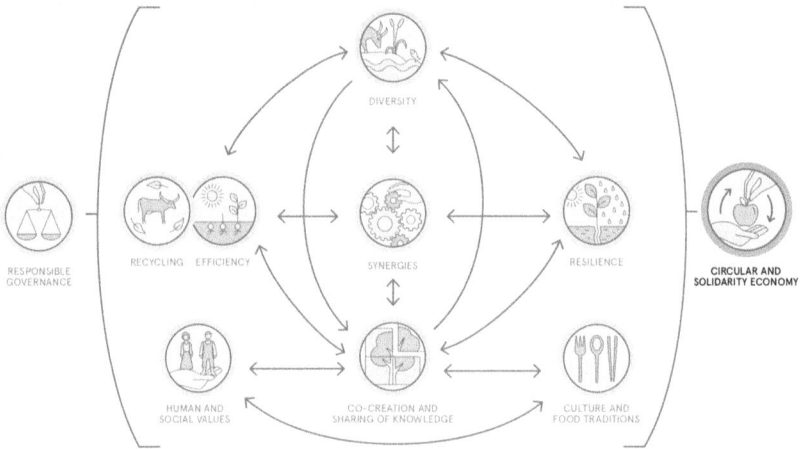

Fig. 2.2 FAO's ten elements of agroecology (*Source*: FAO 2018, The 10 Elements of Agroecology. http://www.fao.org/3/i9037en/i9037en.pdf. Reproduced with permission)

institution, reflecting the UN agency's engagement with civil society in different regions. However, its elements do not centre the political in the same way that authors in NGOs or social movements have—an inevitable result of constraining political processes within FAO itself (e.g. how the organization has historically favoured green revolution-style processes over agroecology; see McKeon 2014).

Moreover, while FAO has incorporated important interconnected issues such as agency and governance in its overall model, those with other agendas might choose elements from it selectively. A few might be 'cherrypicked', for instance, to superficially evaluate an initiative that reduces industrial-chemical usage but does nothing to improve the overall resiliency and integration of the farming system or pays no attention to issues of knowledge, control and power that are critical for agroecology. If agroecology is not based on a shift in power away from elite actors and towards the agency of food producers and strengthening of democracy, it can easily devolve into a technical fix with little potential for wider transformation.

Perhaps the most politically oriented and transformative set of principles that we have encountered are the principles embedded in the Declaration of the International Forum on Agroecology (Nyeleni 2015). These emerged when a number of social movements—including, for instance, La Via Campesina and the World Forum of Fisher People—from all regions of the world came together to articulate an understanding of agroecology based on the principles of food sovereignty and rooted in the voices and priorities of marginalized food producers.

Indeed, many social movements, scientists and governments closely link agroecology to the idea of food sovereignty. They base agroecology in the affirmation of the right to food, the rights of peasants and their cultures, and the fundamental role of food producers and citizens as agents in food practice and policy (Nyeleni 2015; De Schutter 2011). Thus, agroecology is increasingly seen as a practice, a science or a movement (Wezel et al. 2009), or all of these at once (Rivera Ferre 2018).

UN's latest High Level Panel of Experts (HLPE) report, *Agroecological and Other Innovative Approaches* (HLPE 2019), for instance, is an important institutional milestone for agroecology, crystalizing the growing acknowledgement that the system is grounded in the social, cultural and political. The HLPE recognized deficiencies in the FAO ten-element model, added others related to resilience and social equity/responsibility, and introduced "agency" and "ecological footprint" as important concepts in the evaluation of sustainable food systems that enhance food security and nutrition.

In its report, the HLPE (2019) defines agroecology as,

> approaches that favour the use of natural processes, limit the use of purchased inputs, promote closed cycles with minimal negative externalities and stress the importance of local knowledge and participatory processes that develop knowledge and practice through experience, as well as more conventional scientific methods, and address social inequalities. Agroecological approaches recognize that agrifood systems are coupled social–ecological systems from food production to consumption and involve science, practice and a social movement, as well as their holistic integration, to address [food and nutritional security] (p. 14).

Although overlaps exist, the processes and principles underpinning agroecology differ from those in more technology-oriented approaches to sustainable food production (Anderson et al. 2019; Pimbert 2015;

Nyeleni 2015). For instance, climate-smart agriculture, sustainable intensification, some forms of organic agriculture and integrated pest management are all currently in use to frame agricultural transitions, yet generally emphasize technical aspects rather than the political, social and cultural dimensions needed for the transformations needed to address the multitude of crises in food systems today (Pimbert 2015). The collective autonomy and empowerment of food producers lie at the heart of agroecology: that is, local and traditional knowledge, collective action and linkages with consumers and a re-territorialization (see Chap. 11 for 'territorial governance of agroecology transformations') and democratization of food systems.

Governance, power and democracy are central to this vision and practice (González de Molina et al. 2019). By governance, we refer to the dynamics of power, relationships, responsibility and accountability. It is the set of political, social, economic and administrative systems, rules and processes that determine the way decisions are taken and implemented by actors from individuals to institutions and through which decision-makers are held accountable.

With the evolution of the growing number of reports, models, principles and initiatives, agroecology has reports, models and movements, agroecology has grown from a relatively obscure notion to an approach increasingly favoured by policy-makers, intergovernmental organizations, global social movements and the research community. The growing realization of the potential of agroecology has sparked the development of new research centres and a growing number of large-scale research projects, special issues in journals, papers and books. Online case studies, policy analyses, videos and other resources for agroecologists are proliferating. Meanwhile, social movements such as the international farmers' organization Via Campesina have helped to advance a political agroecology. Between 2014 and 2018, FAO organized an intensive global dialogue on agroecology that brought together more than 1400 participants from 170 countries for six regional symposia, taking the debate to a new level and creating many more allies, in governments and elsewhere.

But not all have welcomed agroecology's new prominence or its inclusion in high-stakes governance processes. Proponents of industrial and corporate agriculture view the system with distrust, seeing either a target for co-optation (i.e. depoliticization and watering down) or a threat that must be neutralized (as in recent attacks by the US government on the uptake of agroecology within FAO—see Chap. 10 for 'suppressing agroecology'). These dynamics—of the emergence of agroecology as an alternative paradigm and the relationship with the dominant regime reaction of the dominant regime—are the focus of this book.

Before turning to our theoretical framework and our focus on agroecology transformations, we provide a brief overview of the evidence on the multiple benefits of an agroecological approach.

MULTIFUNCTIONAL BENEFITS OF AGROECOLOGY

Agroecology offers many benefits, from improving yield and profitability to enhancing biodiversity, addressing climate mitigation and providing nutrition. Although it is beyond the scope of this book to fully review the evidence on the multifunctional benefits of agroecology, we provide a short overview and some key studies. As a system that minimizes expensive external inputs and maximizes farm- and community-generated inputs, it is also a boon for the rural poor. A growing body of research indicates that—when appropriately supported and in the right economic conditions—it can outperform conventional systems of agricultural production in many contexts (Pretty et al. 2003; Ponisio et al. 2015).

In a recent meta-analysis, Raffaele d'Annolfo et al. (2017) found that yields increased in 61% of the cases analysed and decreased in 20% while farm profitability increased in 66% of the cases (Betancourt 2020). In another meta-analysis of 118 studies, Ponisio et al. (2015) found that the diversification practices used in agroecological practices can reduce or eliminate any yield gap between organic and conventional agriculture.

Such findings are not limited to the global south. Jan Douwe van der Ploeg et al. (2019) found that in European countries—including the Netherlands, Portugal and Poland—agroecology not only allows for higher yields than conventional systems but also creates employment and considerably improves farmers' incomes as well as the total income generated by the agricultural sector at regional and national levels.

Yet, the idea that alternative agriculture can 'feed the world' is hotly debated. Can it match the yields of industrial agriculture? And might lower yields ultimately lead to further expansion of agriculture and environmental destruction (Kremen 2015)? There is little evidence that high-input industrial systems greatly outperform agroecological practices (Ponisio and Ehrlich 2016). Yet, this narrative—that only high-input farming can feed the world—is often promulgated by proponents of industrial agriculture (see 'feed the world' frame in Chap. 9). Without rehashing the ongoing debate in its entirety, several additional, relevant points can be made.

Research into small- and medium-scale farms, for instance, shows that globally small farmers tend to use less highly intensified practices, yet produce 30–53% of the world's calories and a majority of the world's micronutrients on 24–53% of gross agricultural land (Ricciardi et al. 2018; Graeub et al. 2016).

A second point is that the connections between yield, agricultural expansion and environmental destruction are complicated and contingent. This makes it difficult to unpick a direct link between yield and impact on the environment. Finally, it is incontrovertible that the agricultural system we currently have is not 'feeding the world', despite generating much more food than is necessary while also creating many social and ecological ills (or 'externalities') such as environmental degradation and poor nutrition. The incessant drive to increase yields does not decrease hunger, on the whole. It lies more with political shifts in entitlement and rights that determine if and how people are able to nourish themselves.

It is worth noting that when agroecology is evaluated for its multifunctional ecological and social benefits, beyond mere productivity, it often outperforms high-input systems. In fact, a key limitation in this area may not be agroecology's performance but (a) a hostile context within which agroecology is situated (see disabling factors in each of the six domains of transformation) and (b) a lack of high-quality evidence allowing proper assessment and comparison of agricultural systems, particularly from a multifunctional point of view (Ricciardi et al. 2018).

Improving Biodiversity

Multitudes of farmers, pastoralists, fishers, forest dwellers and indigenous peoples in both the global north and south are agroecologists. All use, sustain and improve biodiversity—genetic to ecological—at scales from farm plots to entire landscapes or territories (FAO 2019; Pimbert and Borrini-Feyerabend 2019).

In agroecological practice, biodiversity is effectively harnessed to improve production, for instance through the use of heterogeneous seeds (e.g. landraces) and breeds, methods such as intercropping, mixed farming, agroforesty and agro-silvo-pastoral systems. These practices, in turn, actively improve biological diversity in a number of ways: conserving it through sustainable use, enhancing the multiple benefits of biodiversity (both wild and cultivated) through their choice of genetic material, design of cropping patterns, development of crop and livestock production systems, and land and water management practices. Biodiversity can also be adaptively managed by groups and networks beyond the farm level at a community, regional or territorial level (Pimbert and Borrini-Feyerabend 2019).

Local knowledge about the properties and dynamic roles of biodiversity in agroecological practices is crucial. The recent UN report *The State of the World's Biodiversity for Food and Agriculture* (FAO 2019) strongly emphasizes the immense contribution of knowledge, skills, innovations and practices of food providers, particularly small farmers, to the conservation, development and sustainable use of wild and cultivated biodiversity and related ecosystem functions.

Addressing the Climate Crisis

Agroecological practices constitute a prime example of nature-based solutions for addressing the climate crisis (IPCC 2019), through both mitigation and adaptation. Compared to on-farm emissions and overall greenhouse gas-driven impacts of industrial agriculture Lin et al. (2011) identify three ways in which agroecology can reduce greenhouse gas contributions of agriculture and food systems:

(a) A decrease in the materials used and amounts of greenhouse gases absorbed or emitted based on agricultural crop management choices (also see: Niggli et al. 2008)

(b) A decrease in the fluxes involved in livestock production and pasture management

(c) A reduction in the transportation of agricultural inputs, outputs and products through an increased emphasis on local food systems

Meanwhile, practices such as the use of organic and green manures, intercropping and tree-planting on farms or in hedges boost organic matter in the soil and, in turn, carbon-sequestration capacity (Lin et al. 2011).

Agroecological strategies can also help farmers adapt to climate change. Crop diversification, the maintenance of local genetic diversity, crop-animal integration, organic management of soils, water conservation and agroforestry, for example, can lay the foundations for a system resilient to shocks and stresses (Brescia 2017; HLPE 2019; Morris et al. 2016). Over the past two decades, observations of agricultural performance and recovery after hurricanes, droughts and other extreme climate-related events have revealed that farms with greater biodiversity are more resilient (Mijatović et al. 2013). Agroecological farms are more resilient to natural disasters such as hurricanes than conventional farms when they are embedded in a complex landscape matrix, are high in biodiversity, employ cropping systems with organic matter-rich soils and deploy water conservation and water harvesting techniques (Altieri et al. 2015). Agroecology, especially its emphasis on robust and resilient networks of mutual aid, has been found to play an important role in social recovery processes such as peace-building and collective responses to disaster—an important aspect of responding to climate change (McAllister and Wright 2019; McCune et al. 2019).

Contributing to Good Nutrition

Agroecologists contribute to dietary diversity and nutrition security through providing diverse dietary offerings for both subsistence (home consumption) and local food markets (Pimbert and Lemke 2018). Farmers' agroecological practices create micro-environments on farms and the wider landscape. In this case, different forms of agricultural biodiversity ('cultivated', 'reared' or 'wild') are utilized by different people, in different seasons, and contribute to dietary diversity and resilience.

Indeed, numerous studies have found that diversified farming systems enhance household dietary diversity and nutrition (Jones 2017). For example, Katie Bliss (2017) examined the farming systems of 30 Nicaraguan households. Because of planting a range of crops harvested at different times, more food was available throughout the year. Ana Deaconu et al. (2019) analysed how agroecological farming systems improve nutrition in poor households in Ecuador through providing food for subsistence, through the generation of income and the empowerment of women who are largely responsible for the nutrition and other reproductive dynamics in households and communities. In a survey of 390 households in Mexico, Javier Becerril (2013) found that the body mass index improved in households using the agroecological *milpa* system (intercropping of maize, beans and squash) compared to households using less diversified farming methods. In northern Malawi, studies have shown that legume intercropping, along with a participatory approach sensitive to cultural values and gender equality, enhanced both food and nutritional security (Nyantakyi-Frimpong et al. 2016).

Strengthening Social Relations

The social impact of agroecology is notable, especially when it is under-pinned by collective, community and territorial processes such as the establishment of food policy councils and peasant-to-peasant learning networks/movements or through the construction of cooperative econo-mies of food distribution (such as community-supported agriculture). Coordinated, collective action is a norm in agroecological practice that is driven by local organizations and movements and the social networks they form at different scales (see Chap 7. on the network domain). In turn, the coordination of local efforts at all scales—from farm to water-shed to the broader landscape—tends to strengthen durable bonds of trust and cooperation amongst proponents of agroecology. Longstanding evidence shows that robust social relations between farmers and other actors in territories and supporting collective action can improve farmers' adaptive capacity by providing opportunities for social learning and for developing collective human energies that can be deployed in times of crisis. For example, collectives of agroecological brigades travelled around to repair farms in response to Hurricane Maria (McCune et al. 2019).

Food producers and their organizations can also gain autonomy through the practice of agroecology by exerting their collective power in local, ter-ritorial, national and international social movements. Indeed, it is this vital function—of offering political agency and outcomes for farmers—that deeply differentiates agroecology from depoliticized and technocratic approaches such as climate-smart agriculture (Pimbert 2015). We now turn in detail to the political function and dynamics of agroecology as the basis for the urgent transformation to a more just and sustainable food system.

CONCEPTUALIZING A TRANSFORMATIVE AGROECOLOGY: POLITICAL ECOLOGY, POLITICAL AGROECOLOGY AND FOOD SOVEREIGNTY

We have emphasized how socio-economic power and good governance are key to agrocology and to transformations towards food and agriculture systems that support, rather than degrade, human and environmental health. That proposition is a premise of *political agroecology*. Substantial research has made the case that within agricultural systems the socio-polit-ical and ecological are inseparable. It has also shown the importance of changing social and political arrangements to foster sustainability in such systems—a baseline in all work within political agroecology.

Political agroecology, in essence, is the application of the methods and concepts of political ecology to agroecology (González de Molina et al. 2019). But how do we define political ecology? Paul Robbins's (2011) influential account describes key elements of the field as "viewing ecological systems as power-laden rather than politically inert", "identifying broader systems rather than blaming proximate and local forces" and "taking an explicitly normative approach [in favour of equity and social justice] rather than one that claims the objectivity of disinterest" (p. 13). Accordingly, political agroecology places current agricultural systems in a historical and geographical context to understand the power relations that give rise to their current dynamics. It exposes the power dynamics that prop up agri-food systems that are environmentally destructive, focus predominantly on increasing yields and profits and are implicated in ongoing undernourishment and rural poverty. At the same time, political agroecology emphasizes the important role of social movements in achieving dignified agrarian sustainability and food sovereignty.

POWER AND POLITICAL INERTIA

Political agroecology is based on the recognition that the current state of any agroecosystem reflects the power-laden relationships of different social actors in that system, such as between agribusiness and farmers or between people of different genders or ethnicity, over time. Thus, any change to an ecosystem is likely to have unequal impacts on different members of society.

Whether 'pristine nature' existed at any point in human history or not, we have long since left that point: every ecosystem on earth has been touched by human activity, whether through direct interaction or through the effects of phenomena such as anthropogenic climate change or the drift of synthetic chemicals. Some ecosystems, of course, are in a less viable and desirable state than others. But who gauges the viability or desirability? This question is inherently about power.

Much of political ecologists' work has aimed to make this point clear. For instance, Nancy Lee Peluso and Peter Vandergeest (2001), among others, have pointed out how governments and other entities have deployed processes of conservation, ecological characterization and management of protected areas to determine which residents of a given territory 'belong' there and which must be removed or barred, for the sake of a given ecosystem. Certain groups, such as indigenous peoples, may be seen as guardians of supposedly pristine ecosystems, or even as part of them; while others may be seen as invaders or opportunists (Durand

2019). Thus, the work of peoples who helped create and maintain habitats (e.g. in the Amazon Rainforest) or place-based agroecosystems can become invisible; such groups may even come to be seen as despoilers (Ghimire and Pimbert 1997).

We do not get to choose the history of an ecosystem or agroecosystem: the peoples and power struggles that have been there continue to affect its characteristics and trajectory. But as the late historian Howard Zinn described history in the title of his 1994 political memoir, "You can't be neutral on a moving train." Choosing to take no side amid ongoing power relationships inevitably gives advantage to whichever group is already more socially powerful. Any intervention in an ecosystem will have different impacts on the different groups interacting with it: some will inevitably receive more benefits or fewer harms than others. Thus, understanding power is fundamental to understanding an ecosystem, and all the more pivotal when there is the potential for new interventions: ecological and social impacts are inextricably linked (Robbins 2011).

Within agroecology, the school of thought most closely concerned with these issues has been called "ecological political economy" (Buttel 2003). Frederick Buttel described agroecologists of this ilk as arguing "that radical changes in the political economy of agriculture and the moral economy of research are needed" if the unacceptably high social and environmental costs of industrial agriculture are to be confronted and reduced.

Academically, the roots of political, transformative agroecology can be traced back through at least the 1960s and 1970s in the United States, with the work of ecologist and biomathematician Richard Levins, among others, in outlets such as the magazine *Science for the People*. Alexander Wezel et al. (2009) also note that the Brazilian agronomist José Lutzenberger meshed scientific analysis of the need for a different agriculture with political analysis and vision. Further back, as mentioned above, were the political economic critiques of industrializing agriculture made by Kropotkin, Liebig, Marx and others in the nineteenth century (Sevilla Guzmán 2011).

The project of political agroecology—critiquing the power dynamics that perpetuate an unsustainable, exploitative agricultural system and working towards systemic transformation—places it among longstanding intertwined traditions of critical theory in academia and social movements. A key, persistent question in political agroecology, then, is how governance, power and control define the choices and agency of farmers and other actors in the food system. Transformation to systems that are more sustainable and just requires an understanding of the dynamics of social change and how and why current systems persist. Thus, we see a world system where food surpluses, wastage and obesity-related chronic diseases occur alongside large-scale hunger and malnutrition; where famines occur even when food is technically available; where industrial models of agriculture persist long after it has become clear even to the mainstream that 'business as usual is not an option'.

These pervasive results arise from a specific history, with powerful actors in governments and corporations maintaining their interests at grievous cost to the environment and, particularly, small-scale farmers and labourers within the food system across the world. Political agroecology is thus an approach to mobilizing knowledge that "allows agroecology and food sovereignty to be put into practice, exploiting the knowledge accumulated by Political Ecology and the experience of social movements" (González de Molina et al. 2019, p. 3). This focus on power, governance and social movements is the foundation on which we build our analysis of the transition and transformation processes in the remainder of the book.

REFERENCES

Altieri, M. A. (2018). *Agroecology: The Science of Sustainable Agriculture*. Boca Raton: CRC Press.

Altieri, M. A., Nicholls, C. I., Henao, A., & Lana, M. A. (2015). Agroecology and the Design of Climate Change-Resilient Farming Systems. *Agronomy for Sustainable Development, 35*(3), 869–890.

Anderson, C. R., Bruil, J., Chappell, M. J., Kiss, C., & Pimbert, M. P. (2019). From Transition to Domains of Transformation: Getting to Sustainable and Just Food Systems through Agroecology. *Sustainability, 11*(19), 5272.

Becerril, J. (2013). Agrodiversidad y nutrición en Yucatán: una mirada al mundo maya rural. *Región y sociedad, 25*(53), 123–163.

Betancourt, M. (2020). The Effect of Cuban Agroecology in Mitigating the Metabolic Rift: A Quantitative Approach to Latin American Food Production. *Global Environmental Change, 63*, 1–9.

Bliss, K. (2017). Cultivating Biodiversity: A Farmers View of the Role of Diversity in Agroecosystems. *Biodiversity, 18*(2–3), 102–107.

Brescia, S. (Ed.). (2017). *Fertile Ground: Scaling Agroecology from the Ground Up*. Oakland: Food First.

Buttel, F. (2003) Envisioning the Future Development of Farming in the USA: Agroecology Between Extinction and Multifunctionality. *New Directions in Agroecology Research and Education*. Madison: UW-Madison, pp. 1–14.

D'Annolfo, R., Gemmill-Herren, B., Graeub, B., & Garibaldi, L. A. (2017). A Review of Social and Economic Performance of Agroecology. *International Journal of Agricultural Sustainability, 15*(6), 632–644.

De Schutter, O. (2011). Agroecology and the Right to Food. *Report presented at the 16th Session of the United Nations Human Rights Council [A/HRC/16/49]*, 8.

Deaconu, A., Mercille, G., & Batal, M. (2019). The Agroecological Farmer's Pathways from Agriculture to Nutrition: A Practice-Based Case from Ecuador's Highlands. *Ecology of Food and Nutrition, 58*(2), 142–165.

Doré, T., & Bellon, S. (2019). *Les mondes de l'agroécologies*. Versailles: Editions Quae.

Durand, L. (2019). Power, Identity and Biodiversity Conservation in the Montes Azules Biosphere Reserve, Chiapas, Mexico. *Journal of Political Ecology, 26*(1), 19–37.

FAO. (2018). *The 10 Elements of Agroecology*. Rome: FAO.

FAO. (2019). *The State of the World's Biodiversity for Food and Agriculture*. Rome: FAO.

Foster, J. B. (1999). Marx's Theory of Metabolic Rift: Classical Foundations for Environmental Sociology. *The American Journal of Sociology, 105*(2), 366–405.

Francis, C., Lieblein, G., Gliessman, S., Breland, T. A., Creamer, N., Harwood, R., et al. (2003). Agroecology: The Ecology of Food Systems. *Journal of Sustainable Agriculture, 22*(3), 99–118.

Ghimire, K. B., & Pimbert, M. P. (1997). *Social Change and Conservation*. London: Routledge.

González de Molina, M., Petersen, P. F., Peña, F. G., & Capor, F. R. (2019). *Political Agroecology: Advancing the Transition to Sustainable Food Systems*. Boca Raton: CRC Press.

Graeub, B. E., Chappell, M. J., Wittman, H., Ledermann, S., Kerr, R. B., & Gemmill-Herren, B. (2016). The State of Family Farms in the World. *World Development, 87*, 1–15.

Hernández Xolocotzi, E. (1977). Agroecosistemas de México: Contribución a la enseñanza, la investigación y la divulgación agrícola.

HLPE. (2019). *Agroecological and Other Innovative Approaches for Sustainable Agriculture and Food Systems That Enhance Food Security and Nutrition*. Rome: High Level Panel of Experts on Food Security and Nutrition of the Committee on World Food Security.

IPCC. (2019). *Climate Change and Land: An IPCC Special Report on Climate Change, Desertification, Land Degradation, Sustainable Land Management*.

IPES-Food. (2016). *From Uniformity to Diversity: A Paradigm Shift from Industrial Agriculture to Diversified Agroecological Systems*. International Panel of Experts on Sustainable Food Systems (IPES).

Jones, A. D. (2017). Critical Review of the Emerging Research Evidence on Agricultural Biodiversity, Diet Diversity, and Nutritional Status in Low- and Middle-Income Countries. *Nutrition Reviews, 75*(10), 769–782.

Kremen, C. (2015). Reframing the Land-Sparing/Land-Sharing Debate for Biodiversity Conservation. *Annals of the New York Academy of Sciences, 1355*(1), 52–76.

Kropotkin, P. A. (2015). *The Conquest of Bread*. Priestland, David (This Edition, Using the 1913 text, First Published in Penguin Classics in 2015 ed.). London: Penguin Classics.

Lin, B, B., Chappell, J., Vandermeer, J., et al. 2011. Effects of industrial agriculture on climate change and the mitigation potential of small-scale agro-ecological farms. CAB Reviews: Perspectives in Agriculture, Veterinary Science, Nutrition and Natural Resources 6, 20, 1–18. http://www.cabi.org/cabreviews

McAllister, G., & Wright, J. (2019). Agroecology as a Practice-Based Tool for Peacebuilding in Fragile Environments? Three Stories from Rural Zimbabwe. *Sustainability, 11*(3), 790.

McCune, N., Perfecto, I., Avilés-Vázquez, K., Vázquez-Negrón, J., & Vandermeer, J. (2019). Peasant Balances and Agroecological Scaling in Puerto Rican Coffee Farming. *Agroecology and Sustainable Food Systems, 43*(7-8), 810–826.

McKeon, N. (2014). *Food Security Governance: Empowering Communities, Regulating Corporations.* Routledge.

Mijatović, D., Van Oudenhoven, F., Eyzaguirre, P., & Hodgkin, T. (2013). The Role of Agricultural Biodiversity in Strengthening Resilience to Climate Change: Towards an Analytical Framework. *International Journal of Agricultural Sustainability, 11*(2), 95–107.

Morris, K., Méndez, V., van Zonneveld, M., Gerlicz, A., & Caswell, M. (2016). Agroecology and Climate Change Resilience: In *Smallholder Coffee Agroecosystems of Central America.* Rome: Bioversity International.

Niggli, U., Fliessbach, A., Hepperly, P., & Scialabba, N. (2008). *Low Greenhouse Gas Agriculture: Mitigation and Adaptation Potential of Sustainable Farming Systems.* Rome: FAO.

Nyantakyi-Frimpong, H., Mambulu, F. N., Bezner Kerr, R., Luginaah, I., & Lupafya, E. (2016). Agroecology and Sustainable Food Systems: Participatory Research to Improve Food Security Among HIV-Affected Households in Northern Malawi. *Social Science & Medicine, 164*, 89–99.

Nyeleni. (2015). Declaration of the International Forum for Agroecology. Available: http://www.foodsovereignty.org/forum-agroecology-nyeleni-2015/ [Accessed June 30, 2016].

Peluso, N. L., & Vandergeest, P. (2001). Genealogies of the Political Forest and Customary Rights in Indonesia, Malaysia, and Thailand. *The Journal of Asian Studies, 60*(3), 761–812.

Pimbert, M. P. (2015). Agroecology as an Alternative Vision to Conventional Development and Climate-Smart Agriculture. *Development, 58*(2), 286–298.

Pimbert, M. P., & Borrini-Feyerabend, G. (2019). *Nourishing Life—Territories of Life and Food Sovereignty* (Policy Brief of the ICCA Consortium No. 6): The ICCA Consortium, Centre for Agroecology, Water and Resilience at Coventry University (UK) and CENESTA (Iran).

Pimbert, M. P., & Lemke, S. (2018). Food Environments: Using Agroecology to Enhance Dietary Diversity. *UNSCN News, (43)*, 33–42.

Ponisio, L. C., & Ehrlich, P. R. (2016). Diversification, Yield and a New Agricultural Revolution: Problems and Prospects. *Sustainability, 8*(11), 1118.

Ponisio, L. C., M'Gonigle, L. K., Mace, K. C., Palomino, J., de Valpine, P., & Kremen, C. (2015). Diversification Practices Reduce Organic to Conventional Yield Gap. *Proceedings Biological Sciences, 282*(1799), 20141396.

Pretty, J. N., Morison, J. I. L., & Hine, R. E. (2003). Reducing Food Poverty by Increasing Agricultural Sustainability in Developing Countries. *Agriculture, Ecosystems & Environment, 95*(1), 217–234.

Ricciardi, V., Ramankutty, N., Mehrabi, Z., Jarvis, L., & Chookolingo, B. (2018). How Much of the World's Food Do Smallholders Produce? *Global Food Security, 17*, 64–72.

Rivera Ferre, M. G. (2018). The Resignification Process of Agroecology: Competing Narratives from Governments, Civil Society and Intergovernmental Organizations. *Agroecology and Sustainable Food Systems, 42*(6), 666–685.

Robbins, P. (2011). *Political Ecology: A Critical Introduction* (Vol. 16): John Wiley & Sons.

Sevilla Guzmán, E. (2011). Sobre los orígenes de la agroecología en el pensamiento marxista y libertario. PLURAL. AGRUCO. CDE. NCCR North South.

van der Ploeg, J. D., Barjolle, D., Bruil, J., Brunori, G., Costa Madureira, L. M., Dessein, J., et al. (2019). The Economic Potential of Agroecology: Empirical Evidence from Europe. *Journal of Rural Studies, 71*, 46–61.

Wezel, A., Bellon, S., Doré, T., Francis, C., Vallod, D., & David, C. (2009). Agroecology as a Science, a Movement and a Practice: A Review. *Agronomy for Sustainable Development, 29*(4), 503–515.

White, M. M. (2018). *Freedom farmers: agricultural resistance and the Black freedom movement*: UNC Press Books.

World Economic Forum. (2018). *Innovation with a Purpose: The Role of Technology Innovation in Accelerating Food Systems Transformations*.

CHAPTER 3

Conceptualizing Processes of Agroecological Transformations: From Scaling to Transition to Transformation

Abstract In this chapter, we survey the recent literature that speaks directly to the issue of bringing agroecology to scale. We discuss the shift towards analytical frameworks that consider not only the farm level but rather whole food system transformations. We then introduce the multi-level perspective on sustainability transitions which we adopt for the purpose of this book. Moving beyond the technical analysis often found in research on sustainability 'transitions', our approach thus adopts agency-centric approach to food systems 'transformation'. To do this, we introduce the notion of domains of transformation, which represent discrete areas where the conflict between agroecology and the dominant food regime manifests and where the potential for collective and transformation is transformation is most potent.

Keywords Scaling up • Scaling out • Multi-level perspective • Sustainability transitions • Domains of transformation

In recognition of agroecology's multifunctional benefits and potential as a paradigm for the future of food, researchers, policy-makers and civil society organizations are converging around the theory and practice of scaling this system. They are looking at how food producers might be encouraged to adopt agroecology and, beyond that, at how agroecology can provide the framework for organizing and transforming entire food systems (IPES-Food 2016, 2018; Mier y Terán Giménez Cacho et al. 2018; IAASTD 2009).

© The Author(s) 2021
C. R. Anderson et al., *Agroecology Now!*,
https://doi.org/10.1007/978-3-030-61315-0_3

29

Three dimensions to this process have been identified. In what is often called horizontal scaling out, "ever-greater numbers of families...practice agroecology over ever-larger territories", engaging "more people in the processing, distribution, and consumption of agroecologically produced food" (Mier y Terán Giménez Cacho et al. 2018, p. 3). Others have argued for the importance of scaling up, in which changes that enable agroecology percolate through institutions, policies and law. A third dimension, deepening, involves seeking ever more synergies and improvements to the agroecological system itself. Yet all these dimensions present significant challenges, including asserting the political nature of agroecology in institutional spaces and policies.

One of the most commonly used frameworks for formulating transitions in agroecology is Stephen Gliessman's (2005) five-level approach. The changes specified in these levels, it should be noted, do not generally unfold successively and neatly; there may be substantial ongoing overlap. At level 1, production is made more efficient by reducing the overall use of inputs (such as fertilizer, fuel or pesticides) across all types of farming systems, from conventional to organic. At level 2, external synthetic inputs are replaced with more sustainable ones, such as biofertilizers or organic pest management products, without fundamentally re-organizing the farming system. Mier y Terán Giménez Cacho et al. (2018) found that many successful transitions to agroecological systems on farms start with simple practices focusing on input substitution or incremental integration (e.g. new synergies between parts of a farming system) that produce benefits quickly (such as boosted yields or cost savings). The latter is important because it helps to motivate producers and may lay a path to more complex and extensive transitions. It is key to note, however, that top-down interventions by governments or aid agencies may view this kind of substitution as a final goal. However, on their own, Gliessman's first two levels are unlikely to be transformative.

Level 3 is a step change: rather than minor tweaks to the existing farming system, it involves a redesign of the entire food and fibre production system based on ecological principles and natural processes. At level 3, multiple agroecological production practices (such as intercropping, compost, mixed farming) are reflexively introduced to foster the development of an intentional agroecological system. While such changes are envisaged as taking place at the farm level, they are deeply shaped by the wider context—the political, economic, cultural and social dynamics that help or hinder farmers' capacity to act. Indeed, the complex integrations between components of the farm involved in agroecological redesign are often only possible when farmers are supported by relationships and structures beyond the farm such as territorial food markets, reciprocal labour arrangements with neighbours or wider diverse landscapes that foster insects and other pollinators. These often enable ecological, political and economic viability that would be difficult if not impossible to achieve as an individual farm. Thus, Mier y Terán Giménez Cacho et al. (2018) found that even in iconic case studies of agroecology transitions, level 3 integration is difficult, and currently relatively rare.

At level 4, connections between producers and consumers are strengthened to support the socio-ecological transformation of the food system. Here, the emphasis is on creating new markets for agroecological farm products and promoting solidarity between farms and their non-farming communities. An even deeper and wider transformation of policies, rules, institutions and culture occur at level 5, which focuses on social justice, democracy and other broad shifts.

While Gliessman's framework has been picked up by the Food and Agriculture Organization of the United Nations (FAO) and many researchers and scientists, the dynamics at level 4 and 5 are mostly referred to in general terms: it is rare to see a concrete description of how transformations of this magnitude happen or the underlying power dynamics. Indeed, early academic work on agroecology was largely focused on the agronomic and ecological dimensions at the farm level (levels 1–3). This scope has shifted in the past five years, but there is still a need to better understand the wider social, economic and political processes that influence broader transformations (Fig. 3.1). Our work engages with the explosion of literature on agroecology and food system change to give nuance to these dynamics.

We view food systems transformation as emergent, non-linear, context-specific, messy processes. The dynamics are not dissimilar to those of struggles towards, for instance, gender equality or environmental governance. Change in these arenas rolls out unevenly and uncertainly, with progress, repression, retrenchment, sudden breakthroughs and gradual changes. Progress is ever-evolving and may only be coherent in retrospect.

Thus, a large-scale transformation of food systems is actually *many transformations*, in which policy changes, struggles and networks that should be aligned are not always aligned (see, e.g., Holt Giménez and Shattuck 2011). "Hopeful monstrosities"—high potential but as-yet crude and inefficient performance—often emerge in these processes (Mokyr, J. 1990. The lever of riches: technological creativity and economic progress, New York: Oxford University Press. [Google Scholar], p. 291, cited in Schot and Geels 2008). Such imperfections are inevitable

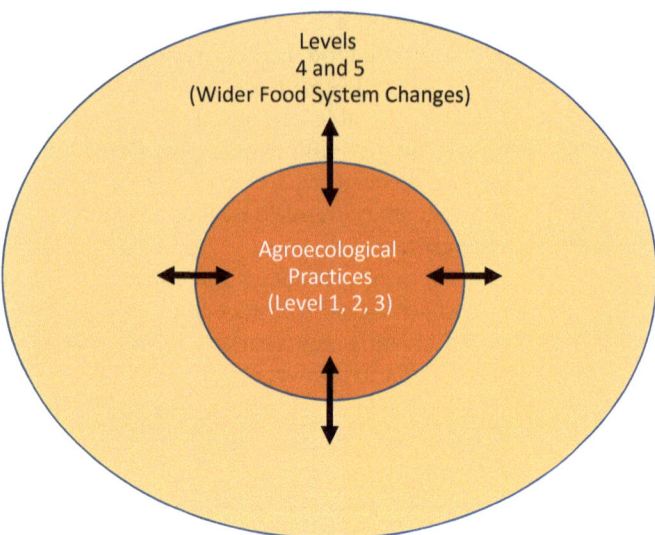

Fig. 3.1 Steve Gliessman's 5-level system has been used to conceptualize agroecology transition. The agronomic emphasis in early agroecology was historically focused on transition at the level of the farm, emphasizing understanding and enabling changes in farm practices (levels 1–3). In recent years, the reconceptualization of agroecology at broader scales and political agroecology as the basis for food-system change has centred analysis on levels 4 and 5. Our analysis in this book focuses on levels 4–5 to interrogate the wider social, political and economic dynamics that underlie the potential for food system transformation and its relationship with agroecological practices

and may even be welcomed as creative experiments needing refinement in the midst of broader emergent transformational processes.

Agroecological transformations are thus not all or nothing. Often, progress can only be made through the kind of retrenchment, 'blowback' or inimical shifts. The process may involve a range of possible pathways, whose direction, speed and scale can be influenced but can never entirely be controlled by individual actors or actions such as state policies. So rather than adopting models such as linear levels of transition, or scaling up, scaling out or transition, we turn in the next section to the *multi-level perspective on sustainability transitions* (MLP) as one suitable, increasingly used framework for studying dialectical and emergent agroecological transformations.

The literature on the MLP is rooted in research on how 'environmental innovations' such as agroecology emerge and unfold. Within that field of study, sustainability transitions are considered "long-term, multi-dimensional and fundamental transformation processes through which established socio-technical systems shift to more sustainable modes of production and consumption" (Markard et al. 2012, p. 956). This tradition examines how environmental innovations have *replaced*, *transformed* or *reconfigured* existing systems.

Its focus is the dynamics, barriers and processes faced in the *transition* towards sustainability. But in this book we choose instead to embrace the idea of *transformation* in the sense of the shifts in power and governance that are central to a political agroecology. Such issues are rarely prominent in the literature on sustainability transitions. And sustainability transitions analysis in the agricultural sector has primarily focused on more technical aspects of the localization of food systems, organic agriculture or permaculture. In contrast, and as we have outlined elsewhere, choose to focus our conceptual framework on the critical issue of agency—that is, on the agency of people working in the fields, in grassroots organizations and social movements. We lift up the politics of the possible, where our analysis goes beyond critique to emphasize political possibility in spite of the grave structural barriers to the agroecology movement's goals (Gibson-Graham 2006) and on frameworks, theories and ideas that strengthen agency and can mobilize action.

Agroecological Transformations Within the Multi-level Perspective

In the literature on sustainability transitions, the multi-level perspective has been used to conceptualize dynamics and patterns in socio-technical transitions—such as in sustainable energy—as "non-linear processes that result from the interplay of developments at three analytical levels". These levels are niches (the locus for 'radical innovations' and the development of alternatives), dominant regimes (the locus of established practices and associated rules that stabilize existing systems) and an exogenous landscape (events and long shifts outside these loci and timeframes, including macro-economic trends, political developments, wars and crises, the ongoing impact of "Deep cultural and societal values, and climate change") (Geels 2011).

Originally, the MLP was developed as a way of understanding how technological innovation can lead to the consolidation of new commercial products in corresponding markets. But agroecology is more than a technological fix; it is an alternative paradigm and a political challenge to the status quo. In light of this, we avoid the technocratic framings of socio-technological regimes (Misra 2017), instead rooting our analysis in political dimensions.

Another body of literature, food regime theory (McMichael 2006), is helpful in the context of the MLP: because its perspective on transitions is explicitly political, it sheds light on global, historic antecedents and political ecological basis of today's dominant food regime. These slowly unfolding landscape-level historical changes (elaborated below in the section on landscape level) set the historical foundations for today's regime (see below section on regime level). Food regime theory also focuses on the semi-stable relationships of power between different actors (e.g. between civil society, government and the private sector); such relationships may also shift during the course of counter-movements (see below section on niche level).

By delving into the history of the current dominant regime, food regime theory has helped to unearth its colonial roots and ongoing dynamics, including racism, euro-centrism, patriarchy and capitalism, currently retrenched through corporate and neoliberal logics.

Landscape Level

In the MLP, the *landscape level* represents macro-scale, often slow-moving contextual factors in society that do not necessarily directly determine change at regime or niche levels but rather "make some actions easier than others" (Geels and Schot 2007, p. 403). Examples of such factors include climate change, demographic change, shifting societal values and macro-economics. In a more specific example, massive shifts in power and sovereignty to corporate actors have been increasingly associated with international trade agreements which then embolden corporation's intellectual property rights, making the spread of corporate-controlled commercial seeds easier and criminalizing many peasant seed networks.

Sudden crises, meanwhile, may often be manifestations of abrupt shifts in such slow-moving processes or by large-scale natural disasters. The worldwide anti-racism protests of 2020, for instance, are a manifestation of longstanding conflict between racial capitalism and the anti-racist counter-movement, ignited in response to the tragic murder of an African-American man, George Floyd, by the white policeman Derek Chauvin in May 2020. Periodic geo-political disasters, such as the food crisis of 2008 and the COVID-19 pandemic, are other recent examples of landscape-level crises that are generally beyond the direct influence of regime or niche actors but can play significant roles—for better or worse—in agroecology transformations. Such crises can be critical in sustainability transitions because they act as catalysts, creating pressure that can destabilize a regime and open up opportunities for alternatives to thrive (Mier y Terán Giménez Cacho et al. 2018). The opposite can, of course, occur: crises may give powerful actors in a regime an opportunity to reify their power and further entrench the status quo. For example, the 'Mad Cow Crisis' in the United Kingdom led to a further centralization and consolidation of power in the meat processing industry and food system, further undermining local food systems.

At the same time, an under-explored area of agroecology transformation is the role of 'sparks' other than crises that can ignite change. For example, John Kingdon's classic 2011 work on political agenda-setting points out several precipitating factors ("focusing events", such as disasters, crises or new discoveries) and opportunities for significant change ("windows", when social and political actors are primed to act for change, such as elections or mass protests spurring society to action) (Kingdon 2011). Other precipitating factors can include dramatic changes in established indicators (e.g. the large increase in hunger and the use of foodbanks seen in recent years under austerity measures in the United States and the United Kingdom), influential new framings (e.g. discursive changes like 'Black Lives Matter' or 'The 1%'—see Chap. 9 on the discourse domain) and the emergence and diffusion of a powerful symbol (the notion of extinction rebellion as a symbol to motivate action on climate change). Nor are crises enough to provoke agroecological transformation without preliminary political and organizational groundwork, as described by Eric Holt-Gimenez in the case of the deadly 1998 Hurricane Mitch. The best time to organize for transformation is before a crisis (or other 'window') manifests.

Food regime theory provides important analysis of broad shifts in the global political economic landscape over time that are particularly relevant for the landscape level. One commonality of all approaches to food regime theory is a central focus on how "forms of capital accumulation in agriculture constitute global power arrangements, as expressed through patterns of circulation of food" (McMichael 2009, p. 140). Food regime theory has remained centred, within the global historical context, on the contradictions and political struggles between different social groups in relation to food and agriculture. Those adopting the theory also generally accept that it focuses on the "rule-governed structure of production and consumption of food on a world scale" (Friedmann 1993, pp. 30–31). And they broadly agree on the existence of two food regimes, identified in early analyses.

The first, a 'colonial-diasporic' regime lasting from 1870 to the 1930s, was shaped by European colonialism and direct expropriation-dispossession from its colonies. The second 'mercantile-industrial' regime, running from the 1950s through the 1970s (and possibly into the present), was fundamentally shaped by the Cold War, the United States and its "informal empire of postcolonial states", international developmentalist programmes and strategic uses of agricultural surpluses (McMichael 2009). As for now, Philip McMichael observes that a third, "possibly emergent", regime has deepened the processes of the second regime. Alongside ongoing consolidation in food supply chains and food retail—this "emerging global food/fuel agricultural complex" is often termed the "corporate food regime" (Friedmann 2006).

Consistent with food regime theory, analysts who suggest that we are seeing a third regime identify it as a significant historical moment fundamentally shaped by tensions between "opposing geo-political principles" (McMichael 2009) in a corporatized "world agriculture". In this regime, power has shifted away from the nation-state whereby a reconfiguration of power through neoliberal globalization and trade has given primacy to the global power of corporations in driving food regimes.

Grasping the dynamics of these landscape-level factors is important. However, it is in the interface between the niche and regime levels that the possibilities for agency are tangible and immediate.

Regime Level

The *regime level* in the MLP represents "established practices and associated rules that stabilise existing systems" (Geels 2011, p. 26). Alignments and interdependencies of laws, processes, infrastructure and regulations across regimes tend to become locked in, building resistance to path-breaking innovations.

The thrust of today's dominant food regime can be characterized by corporatization, a productivist (a reductionist focus on yield and profit) mentality, a reification of racist and patriarchal structural violence (e.g. the racialized and gendered patterns of poorly paid dangerous work around the globe) and an emphasis on monocultural, high-input, energy-intensive agriculture. This regime aims to standardize all aspects of food system to enable industrialization, decrease costs of production and increase profits. This push towards uniformity plays out not only in agronomics (monocultures of seeds, crops and livestock) but also in the erasure of diverse knowledge systems, markets and territorial agroecosystems that, as we will outline, are fundamental for agroecology and for sustainable food systems.

Powerful actors working on behalf of corporate and elite interests (e.g. lobbyists, politicians) also often attempt to resist or appropriate change in order to maintain the status quo. For example, where a government adopts a narrow technical understanding of agroecology into national frameworks, the system is at risk of being co-opted (Ajates Gonzalez et al. 2018). Structural power within a regime privileges particular actors at the expense of others along intersecting dimensions of oppression (e.g. racism, patriarchy, class).

However, this dominant regime has not been adopted universally over time and geographically; it manifests in different places in different ways, at different times, and is always met by resistance. In this context, transformations involve coalitions of actors involved in political, economic, cultural and social struggles and with competing interests, seeking to shape both the regime and emerging alternatives in particular places. A shift in power relations triggers the transformation—in particular, where disenfranchized actors and groups gain agency and power. In this book, we focus on agroecology as one specific form of resistance, which we further elaborate on in the next section as an emerging global niche.

Niche Level: Agroecology and Food Sovereignty

Within (and against) this corporate food regime are social movements that are known as 'niches' within the context of the MLP. McMichael (2009) describes the dominant regime vis-à-vis the agroecological niche as "food from nowhere" (undifferentiated, appropriated and commercialized) and "food from somewhere" (grounded in place, space and culture)—that is, empowering local and traditional producers and food cultures.

In MLP parlance, niches represent the emergence of radical sociotechnical alternatives to dominant principles and ways of working. This distinguishes them from 'market niches'—specialized products, technologies or services within capitalist markets. Agroecology, with its emphasis on principles such as ecological processes, low external inputs, the agency of food producers and consumers, and autonomy from elite and corporate power, sharply contrasts with the incentives, policies, programmes, rules and norms of the dominant regime (Smith and Raven 2012).

In our framework, agroecology as a 'niche' does not only refer to single projects, clusters or localized experiments; it also embraces proto-regimes or counter-regimes that emerge from such local projects and inform them in turn. Within the MLP, such local experiments take place in 'protective space' on farms and in communities, shielded from a potentially hostile dominant regime through, for example, exemptions from regulations. Political agroecology is explicitly constructed to contest, deconstruct, transform and replace the dominant regime and is often born of radical demands for food sovereignty (Nyeleni Movement for Food Sovereignty 2007). Using this interpretation of niche level helps us to move analytically from seeing isolated 'innovations' in local niches as disconnected phenomena to seeing how—even when geographically separated—the experiments and experiences of niches can be a part of dynamics and movements working across space and time in processes of transformation.

As the concerns and positionality of political agroecology can be seen reflected in the demands of food sovereignty, let us look at those demands, as laid out in the 2007 Nyeleni Declaration:

1. A focus on food for people, with rights to sufficient, healthy and culturally appropriate food at the centre of food policies rejecting the treatment of food as just another commodity produced for the purpose of profit and the concomitant immorality of access to food depending on economic resources
2. Valuing food providers, particularly with regard to securing rights and respect for those who grow, harvest and process most of the world's food: farmers and workers within small-scale, family, traditional and indigenous food systems
3. The localization of food systems, *inter alia*, in contrast to the currents of capital favouring large corporations
4. Local control of food providers and consumers over territory, land, grazing, water, seeds, livestock and fisheries based on the rights of local food producers and inhabitants in territory—food sovereignty rejects the privatization of such resources, for example through intellectual property rights regimes or commercial contracts
5. Broad-based skill-building that supports indigenous knowledge in local communities, in part through the management, conservation and development of local food production, harvesting and distribution systems and appropriate research supporting these activities
6. Working with nature by respecting and supporting the integrity and contributions of ecosystems and communities' ecological knowledge, particularly the use of diversified agricultural methods reliant on few external inputs

In Canada, a seventh principle was added by the Indigenous Circle of the People's Food Policy:

7. Food is sacred and not to be squandered.

This final demand counters the commodification of food and claims that the spiritual and cultural dimensions of food are fundamentally important: food is central to who we are as people. It reflects an indigenous cosmovision based on respect towards nature.

Political agroecology, as the conceptual basis for putting agroecology and food sovereignty into practice, is not the only possible formulation of such a niche. Others have claimed organic, sustainable or climate-smart agriculture, alternative food networks and other systems as alternatives. While related to agroecology, these systems are not always clearly allied to political transformation; they often leave existing power dynamics in place and in many ways reinforce the dominant regime and the position of actors in it. Some approaches, such as organic agriculture, had a radical and transformative agenda that, over time, has been twisted to conform to the dominant regime in many respects (even if a radical movement does continue to exist alongside the more mainstream dimensions of organic agriculture and markets). As a radical alternative, agroecology is thus also competing with niches that are much more aligned to and supported by the dominant regime.

While agroecology and food sovereignty are not immune from being co-opted or deployed with a view of superficial reforms (as will be discussed later in the book), they are currently the most significant, well-developed and coherent formulations for advancing a counter-regime. Within the wider context of the world historical analysis provided by food regime theory, it is in this active formulation of political agroecology that social movements are pushing for emancipatory and transformative change—and that the social agency of affected peoples is realized from the bottom up.

Our Approach: Advancing an Agency-Centric Approach to the Governance of Agroecology

In the MLP, transition is driven by interactions between the three levels, especially—as we have mentioned—niche and regime. While niches may influence the regime, the regime may act or react in ways that affect the niche's growth. Generally speaking, as we have shown, regimes are configured to maintain the status quo and therefore may marginalize or co-opt emerging alternatives while actors within the regime work politically to maintain this position. This does not mean that socio-technological configurations such as regulations, policies and laws are fixed or that dominant regime actors want them to remain unchanged. On the contrary: changing, rearranging and tweaking rules, norms, technologies or laws often reinforce their power and position.

As such, issues of power and governance are critical but often underappreciated in the understanding of sustainability transitions through the MLP. Governance is the result of numerous interactions among actors in the government, private sector and civil society who, directly or indirectly, shape its content, interpretation and implementation. Omar Felipe Giraldo and Peter Rosset (2018) use the term "territory in dispute" for this battle in governance between institutions in the regime and social movements advancing agroecology as political struggle. This struggle plays out in economies, the environment and other material dimensions but also in the realm of ideas. Both are central to our analysis.

In examining how power and governance might be shifted to allow agroecology to fulfil its emancipatory potential, some questions can be helpful. Which actors are involved? Where does 'governance-making' actually take place? Who has final control over decision-making processes? Whose perspectives, knowledge, values and aspirations are embedded in governance and whose are excluded? Through which avenues can governance be improved? Whose interests are served, and is someone held accountable? Asking these questions helps to shift attention from an analysis of governance per se to the analysis of the governance *process*. This is important because, given the often-contested pathways and goals of agroecological transformation, governance must be equitable.

Wider political, economic, social, cultural and ethical contexts—as well as norms and rules of power—shape any governance system for agroecological transitions. The architecture of such systems covers a spectrum, from highly centralized, uniform, top-down and coercive decision-making to decentralized, horizontally distributed, participatory forms that are socially inclusive and tailored to local contexts (Pimbert 2018). Some prevailing governmental systems are highly prejudicial, excluding, disempowering and oppressing groups including women, people of colour, ethnic minorities, indigenous peoples, LGBTQ+ groups and youths. Thus, when we refer to confronting an unjust and unsustainable dominant regime, we are also referring to its systemic racism and patriarchal, heteronormative stance (see especially Chap. 8 on equity).

It is clear that corporate regimes lock in many unjust aspects of governance of food and agriculture and that these ultimately influence all emerging alternatives or attempts for autonomy. In our view, this is inescapable. Any notion of building completely autonomous alternatives is naïve, at base. Moreover, engaging in transformative action at the niche-regime interface involves a fraught engagement with the power of the regime, which often operates in subtle, hidden and systemic ways (Laforge et al. 2016). Collective efforts to shift power thus demands care and awareness of any implications of various actions and interventions.

While the tensions, contradictions and risks are very real in these governance interventions, so too are the possibilities for transformation. Even agroecology niches that flourish on their own terms may not contribute to transformation if the regime and its governance are not challenged. The parts on the domains of transformation (Part II) and on the six 'effects' of governance interventions (Part III, chapter 10) offer insights into how this can be accomplished.

Finally, a word about scale in relation to governance. Complex forms of governance occur across multiple interacting scales, and the enactment of power at one scale is influenced by the relationship between actors and institutions at others—household, local, territorial, national and international. For instance, international trade agreements, such as the North American Free Trade Agreement (NAFTA), will shape possibilities for developing alternative food networks in specific territories and will also have implications for individual households. While a deep dive into these dynamics is beyond the scope of this book, we will share insights on them in examples.

In part III we present the territorial scale as decisive in food systems transformation. Defined by a range of spatial, environmental, political and cultural factors, the territory is a key meeting place for actors, sectors, ideas and practices and for the interaction of food producers' strategies with state policies. But any efforts to enable a territorial approach also demands work at other levels of governance—in part, to transform relationships and structures at those scales (e.g. global) where political power has congealed.

Introducing Domains of Transformation

To better understand and construct an agency-centred approach to agroecology transformations, we have developed the notion of the *domains of transformation*. These domains represent discrete spheres of activity within which agroecology (a 'niche' in MLP terms) and the dominant regime come into conflict. Parsing out these niche-regime interactions into domains helps to break down the transformation pathway into key areas of intervention. However, it is vitally important to understand that these domains are interrelated and should not be viewed in isolation. An intervention in one domain often has an effect in another domain, as we will illustrate below. Thus, the overlap and alignment of domains is critical in accelerating agroecological transformation, which calls for a holistic and cross-domain analysis and transformation strategy.

We derived these domains through a process of collectively analysing the growing literature and case studies on agroecology transitions (see Anderson et al. 2019 for a more detailed methodology). Through an iterative process of collective reading, group discussion and diagramming our emerging analysis across these studies, we developed our framework of six domains of transformation. We started by identifying the enabling factors and disabling factors that we found in each study. As we viewed the emerging patterns across the studies it became clear that there were six main domains within which the majority of the enabling and disabling factors could be attributed. It also became clear that these domains were situated at the intersection of and conflict between the niche—where proponents of agroecology are strategically working to enable agroecology—and the regime—where inertia of the dominant system and resistance by dominant actors disable agroecology (Figs. 3.2 and 3.3).

Fig. 3.2 Within each domain, there are factors, dynamics, structures and processes that constrain agroecology (orange examines these dynamics within six arrows), and those that enable it (blue arrows). Our analysis shows interdependent domains of transformation (see Fig. 3.4)

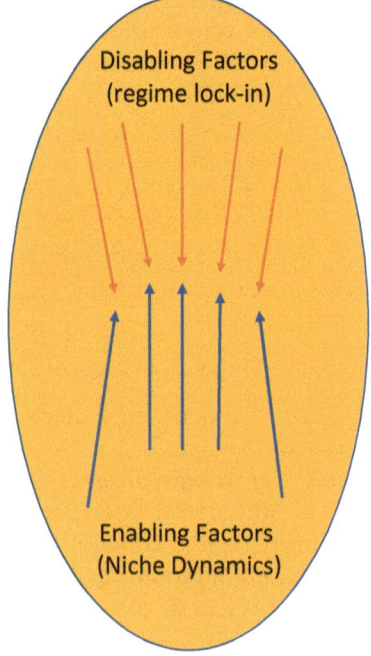

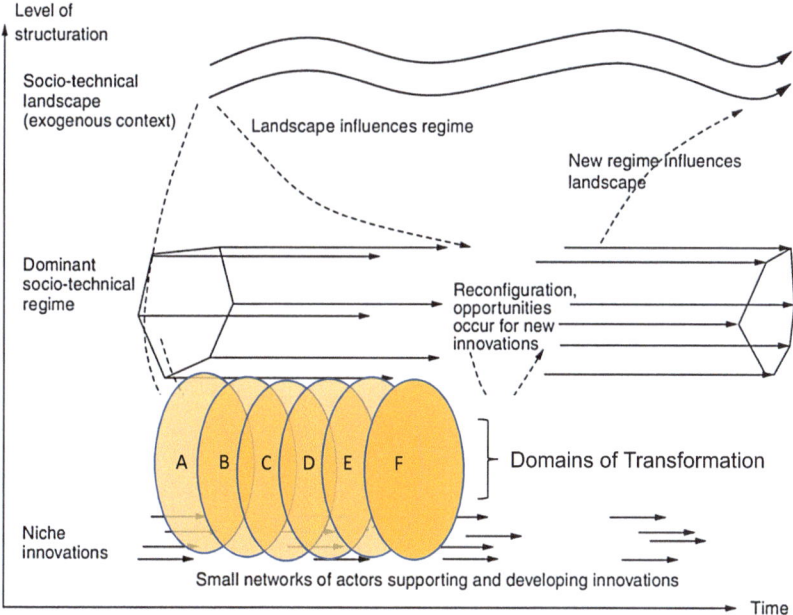

Fig. 3.3 Domains of transformation are depicted here as definable interfaces between niche and regime superimposed onto a simplified version of Frank Geels and Johan Schot's (2007) multi-level perspective figure

The six domains (Fig. 3.4) that are critical in agroecological transformations are: rights and access to nature, knowledge and culture, systems of economic exchange, networks, equity and discourse. In Part II, we analyse the enabling and disabling dynamics within each of these domains. We will then go on to discuss in Part III how, in these domains, transformations in governance and power relationships can gain strength, gradually enabling agroecology—a new, more just food regime—to take root.

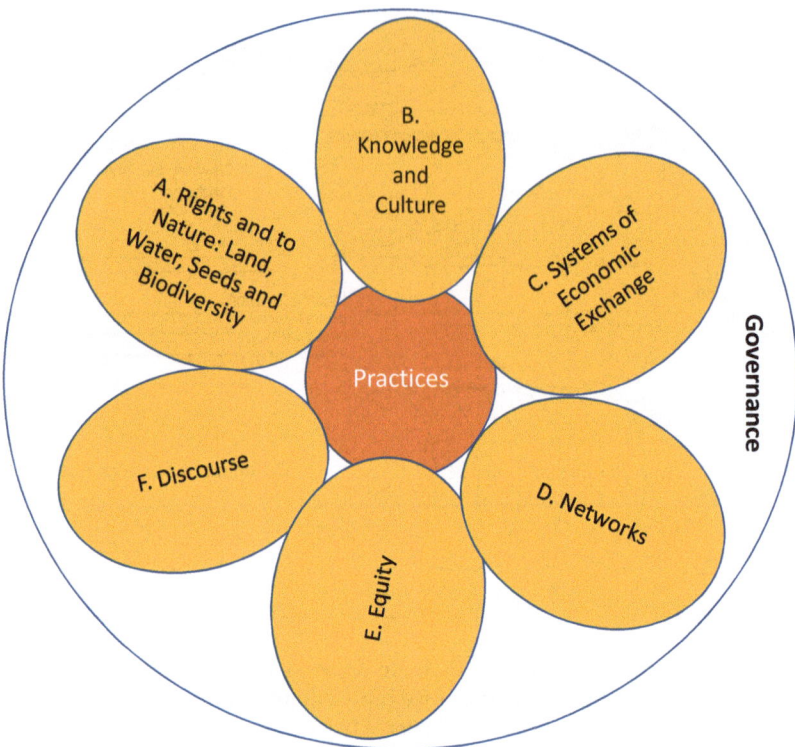

Fig. 3.4 These six domains of transformation, within which agroecology comes into conflict with the dominant corporate food regime, are critical sites of intervention in pursuit of agroecology transformations. The extent and depth of agroecological practices on farms and in territories are shaped by processes of governance, power and control as they manifest in and across these domains

REFERENCES

Ajates Gonzalez, R., Thomas, J., & Chang, M. (2018). Translating Agroecology into Policy: The Case of France and the United Kingdom. *Sustainability, 10*(8), 1–19.

Anderson, C.R., Bruil, J., Chappell, M.J., Kiss, C., and Pimbert, M.P. (2019). From Transition to Domains of Transformation: Getting to Sustainable and Just Food Systems through Agroecology. *Sustainability 11*(19). https://doi.org/10.3390/su11195272.

Friedmann, H. (1993). The Political Economy of Food: A Global Crisis. *New Left Review, 197*, 29–57.

Friedmann, H. (2006). From Colonialism to Green Capitalism: Social Movements and Emergence of Food Regimes. In F. H. Buttel & P. McMichael (Eds.), *New Directions in the Sociology of Global Development* (Vol. 11, pp. 227–264, Vol. (Research in Rural Sociology and Development)): Emerald.

Geels, F. W. (2010). Ontologies, socio-technical transitions (to sustainability), and the multi-level perspective. Research Policy 39: 495–510.

Geels, F. W. (2011). The Multi-level Perspective on Sustainability Transitions: Responses to Seven Criticisms. *Environmental Innovation and Societal Transitions, 1*(1), 24–40.

Geels, F. W., & Schot, J. (2007). Typology of Sociotechnical Transition Pathways. *Research Policy, 36*(3), 399–417.

Gibson-Graham, J. K. (2006). *A Postcapitalist Politics*. Minnesota: University of Minnesota Press.

Giraldo, O. F., & Rosset, P. M. (2018). Agroecology as a Territory in Dispute: Between Institutionality and Social Movements. *The Journal of Peasant Studies, 45*(3), 545–564.

Holt Giménez, E., & Shattuck, A. (2011). Food Crises, Food Regimes and Food Movements: Rumblings of Reform or Tides of Transformation? *Journal of Peasant Studies, 38*(1), 109–144.

IAASTD. (2009). *Agriculture at a Crossroads: Report of the International Assessment of Agricultural Knowledge, Science, and Technology*. Library of Congress.

IPES-Food. (2016). *From Uniformity to Diversity: A Paradigm Shift from Industrial Agriculture to Diversified Agroecological Systems*. International Panel of Experts on Sustainable Food Systems (IPES).

IPES-Food. (2018). *Breaking Away from Industrial Food and Farming Systems: Seven Case Studies of Agroecological Transition*.

Kingdon, J. W. (2011). *Agendas, Alternatives, and Public Policies* (2nd ed.). New York: Harper Collins.

Laforge, J. M. L., Anderson, C. R., & McLachlan, S. M. (2016). Governments, Grassroots, and the Struggle for Local Food Systems: Containing, Coopting, Contesting and Collaborating. *Agriculture and Human Values, 34*(3), 663–681.

Markard, J., Raven, R., & Truffer, B. (2012). Sustainability Transitions: An Emerging Field of Research and Its Prospects. *Research Policy, 41*(6), 955–967.

McMichael, P. (2006). Global Development and the Corporate Food Regime. *Research in Rural Sociology and Development, 11*, 265–299.

McMichael, P. (2009). A Food Regime Genealogy. *Journal of Peasant Studies, 36*(1), 139–169.

Mier y Terán Giménez Cacho, M., Giraldo, O. F., Aldasoro, M., Morales, H., Ferguson, B. G., Rosset, P., et al. (2018). Bringing Agroecology to Scale: Key Drivers and Emblematic Cases. *Agroecology and Sustainable Food Systems, 42*(6), 637–665.

Misra, M. (2017). Moving Away from Technocratic Framing: Agroecology and Food Sovereignty as Possible Alternatives to Alleviate Rural Malnutrition in Bangladesh. *Agriculture and Human Values, 35*(2), 473–487.

Nyeleni Movement for Food Sovereignty. (2007). *Nyéléni Declaration for Food Sovereignty.*

Pimbert, M. P. (2018). *Food Sovereignty, Agroecology and Biocultural Diversity: Constructing and Contesting Knowledge.* London: Routledge.

Schot, J., & Geels, F. W. (2008). Strategic Niche Management and Sustainable Innovation Journeys: Theory, Findings, Research Agenda, and Policy. *Technology Analysis & Strategic Management, 20*(5), 537–554.

Smith, A., & Raven, R. (2012). What Is Protective Space? Reconsidering Niches in Transitions to Sustainability. *Research Policy, 41*(6), 1025–1036.

Steve Gliessman (2016). Transforming food systems with agroecology, Agroecology and Sustainable Food Systems, 40:3, 187–189, https://doi.org/10.108 0/21683565.2015.1130765

PART II

Domains of Agroecology Transformations

CHAPTER 4

Domain A: Rights and Access to Nature— Land, Water, Seeds and Biodiversity

Abstract This chapter discusses a seemingly obvious but often underappreciated reality—without secure land tenure as well as access to and control over other elements of natural ecosystems, agroecology specifically, and the sustainable livelihoods of food producers more generally, will be impossible. We review how the access and control over water, ecosystem, cultivated biodiversity, seeds, breeds and soil amongst other aspects of nature enable agroecology. Conversely, we review the disabling conditions in this domain where inadequate and insecure access and tenure rights for various elements of natural ecosystems increase vulnerability, hunger and poverty and undermine agroecology. Insecure rights and access to nature provides little incentive for farmers, communities and territorial networks to invest in long-term agroecological approaches.

Keywords Land tenure • Seed systems • Rights • Land reform

Equitable and secure access to, and control over, rights to land and forests, water and fisheries, and seeds and biodiversity are referred to collectively in this book as 'nature'. They are essential to agroecology transformations in every territory. If people do not feel secure about their long-term right to nature, there is little hope for a viable agroecology or for improving well-being and sustainable livelihoods. We examine this domain in this chapter.

C. R. Anderson et al., *Agroecology Now!*,
https://doi.org/10.1007/978-3-030-61315-0_4

Societies define and regulate the way food producers gain control over and access to land, fisheries and forests through formal and informal tenure systems. These systems determine who can use which resources when, for how long and under what conditions. They may be based on written policies and laws or on unwritten customs and practices (Ostrom 1990) and cover a spectrum from communal to private (Box 4.1).

Secure land tenure and land reform, along with socially protected access to nature, have long shown to be vitally important for smallholder livelihoods, culture, well-being and investment in sustainable agriculture, including agroecology (Lawry et al. 2016). Because subsistence activities depend on nature-based cultural practices, this domain is particularly important for indigenous peoples.

Inadequate or insecure tenure rights to ecosystems, however, may lead to vulnerability, hunger, poverty, conflict and environmental degradation. Insecure tenure offers little incentive for farmers, communities and territorial networks (e.g., farmers' unions and rural social movements) to invest in agroecology for a long term. For indigenous communities in particular, Ellen Woodley et al. (2006) argue that privatization or concessions "by governments or even by Indigenous Peoples themselves to commercial enterprises for logging, mineral and oil exploitation, hydro-electric dams, plantations or designation as national parks frequently destroys their traditional food and agroecological systems and their cultural identity". Woodley adds that changes emerging from such developments in indigenous areas, such as "forced resettlement, compensation, registration of household heads for taxation" or work in extractive industries have not favoured women, eroding their rights, prosperity and status.

Box 4.1   Tenure Arrangements: From Nature Privatized to Nature as Commons (and Beyond)

The classic view of sustainable and productive use of natural resources insisted that the only 'solutions' to stop overexploitation and collapse were two: either privatize them so that owners have an incentive to care for their own 'piece' of the resources or place them under government regulation to restrict their use to sustainable levels (cf. "mutual coercion mutually agreed upon by the majority of the people affected": Hardin 1968). Elinor Ostrom (1990, p. 13) observed that many thinkers have insisted on privatization as the "only" solution.

(continued)

Box 4.1 (continued)

However, analysts have pointed out multiple problems with privatization. For one, it does not address the power differences between those with different abilities to purchase, maintain and defend their ownership (Robbins 2011). Nor does privatization address the equity implications and inequalities that often accompanied "original" ownership claims ("primitive accumulation") or ongoing processes of exploitation and expropriation, "accumulation by dispossession" (Araghi 2008; Robbins 2011). At the same time, privatization and ownership are not any single 'thing' but rather packages of rights (such as access, exclusion, use, disposal). José Vivero-Pol et al. (2018) have critiqued the creep of commodification and privatization in agri-food systems and natural resource systems more broadly, pointing out the necessary diversity of arrangements and negotiations to sustainably and equitably manage natural resources.

The literature on 'the commons'—resources accessed and used by many individuals in common—offers examples of communal ownership and rights that can support agroecology. Insights from such studies and traditional knowledge highlight nuanced, dynamic and sustainable systems of access for land, water, seeds and trees (Ostrom 1990). Analysts have increasingly argued that food and agriculture systems generally should be viewed as systems of commons (Vivero-Pol et al. 2018).

Enabling Conditions

Land

Much evidence points to secure land tenure as encouraging agroecological approaches and having positive impacts on environmental sustainability, efficiency, equality, productivity, income stability, poverty and hunger reduction (Deininger et al. 2009; Lipton 2009; Higgins et al. 2018). Studies have also revealed causal connections between secure land tenure and investment in agricultural goods and methods that may only bear fruit in the long term, such as planting trees as crops, maintaining and improving soil quality, and preventing erosion and other forms of degradation (Fraser 2004; Otsuka 2001).

Land reform—changing laws, regulations or customs regarding land access and ownership—enables secure land tenure. And land reform and agroecological approaches have also proven to be mutually reinforcing in at least some cases, with land reform enabling locally tailored approaches to agroecology and agroecology in turn supporting more sustainable, improved livelihoods for farmers.

Two decades ago, development economist Michael Lipton and colleagues noted that redistributive land reform was a long-established, yet undervalued, policy approach (Lipton et al. 1998, p. 112). Subsequently, in 2009, Lipton noted that land reform—which he defined as "laws with the main goal of reducing poverty by substantially increasing the proportion of farmland controlled by the poor, and thereby their income, power or status"—remained both key and underused in policy. Theoretical and empirical evidence, he wrote, pointed to its positive effects on environmental sustainability, efficiency, equality, productivity and income stability, as well as reducing poverty and hunger.

Globally, experience confirms this statement. In Brazil, what some analysts have called 'ecological' land reform has taken place since the 1980s, with movements like the MST (Landless Rural Workers' Movement) occupying land to support production for local and national consumption rather than export. For example, many MST settlements have focused on growing locally eaten staples rather than crops that do not contribute directly to subsistence, like sugarcane. At the same time, they also work as partners in protection and reforestation of bordering protected forest areas. In other words, many participants in land reform have worked to incorporate social and environmental goals (conservation, subsistence, sufficient income and dignified livelihoods) into planning and implementation of the new settlements resulting from land reform (Chappell et al. 2013; Wittman 2010). Efforts such as those by movement settlers in 14 studied settlements in Mato Grosso to work together to transition to agroecological production and protect forest and river reserves are not a norm in Brazilian land reform settlements. Yet, surveys of about 1500 settlers across 92 settlements in 2000 and 2001 found evidence of increased crop diversity, improved food security and increased self-reported quality of life (Heredia et al. 2016).

Further evidence of enabling dynamics between land reform and agroecology can be seen in the founding of the Latin American Institute of

Agroecology at an MST settlement (an area where MST members have staked a claim and/or successfully secured redistributed land) and the increased focus on agroecology education within the MST (Schwendler and Thompson 2017). Although agroecology seems a relatively small but growing focus within the landless movement,[1] land reform does seem to be a crucial enabling factor for agroecology, and vice versa.

Land reform has, however, taken many forms across time and location, and not all have enabled the development of agroecology. Jun Borras (2007) found that the only system bettering the lives of agricultural producers was one that effectively redistributed socio-political power and resources—which is also a bedrock demand of political agroecology and food sovereignty, reemphasizing how land reform and agroecology can be mutually reinforcing. Borras observed that land redistribution actually exists as a spectrum—from arrangements where poor and landless farmers are given access to new land but remain essentially as tenant farmers as they have to pay off the government or previous owner to arrangements where land is expropriated from wealthy landowners with limited or no compensation and the formerly landless are immediately granted full ownership as well as social support to start up production on their new land. Correspondingly, outcomes for agroecology and farmers' well-being and prosperity are always partial and contingent. For example, a systematic review of land reform and its impacts found that secure tenure had a positive impact on "productive and environmentally-beneficial agricultural investments" such as agroecology, as well as empowerment of women, but did not support "links with productivity, access to credit, and income" (Higgins et al. 2018).

The Food and Agriculture Organization of the United Nations' (FAO) Voluntary Guidelines on the Responsible Governance of Tenure of Land, Fisheries and Forests in the Context of National Food Security, also known as VGGT, provides promising guidelines on how to navigate complex land reform processes in ways that maintain or increase social justice. Notably, it emphasizes the importance of recognizing and protecting "all legitimate tenure rights, including customary tenure systems and legitimate customary tenure rights that are not currently protected by law" (Civil Society Mechanism for Relations to the Committee on World Food Security

---

[1] The landless movement involves groups of displaced farmers, rural labourers without their own land, dispossessed indigenous peoples and others attempting to get fairer access to the huge swathes of land in the hands of a small number of wealthy owners.

(CSM) 2016). However, customary tenure systems can themselves leave in place inequalities in social divisions such as gender (Collins 2014). Accordingly, the VGGT suggest multiple pathways for addressing gender inequality, and other issues, through continuous and participatory processes.

In 2012, the Committee on World Food Security (CFS), an intergovernmental and intersectoral forum within the United Nations' system, endorsed the VGGT. Although the guidelines are an international instrument, they have been used extensively within individual countries and subnational territories and communities as well as at the global and regional level (Duncan 2015; Civil Society Mechanism for Relations to the Committee on World Food Security (CSM) 2016).

Complementary principles and parallel suggestions for land tenure governance can be found in documents such as the Voluntary Guidelines for Mainstreaming Biodiversity into Policies, Programmes and National and Regional Plans of Action on Nutrition; for Conservation and Sustainable Use of Crop Wild Relatives and Wild Food Plants; for Seed Policy Formulation; and still other voluntary guidelines around small-scale fisheries and soil. There are as yet fewer academic and civil society analyses of these documents, however, than there are for the VGGT.

Water

For farmers adopting agroecology, access to water is obviously essential—for production and for the transformation and preparation of food and fibre within territories. Securing local rights to access, use and control supplies of water also enhances, in turn, local communities' capacity and resources for monitoring and maintaining their broader rights (e.g. "substantive citizenship", as presented by Ribot 2014).

There are many factors determining access to and control over water: property rights, social and political institutions, and cultural and gender norms; the way water is managed, priced and regulated in water basins and at the local level also plays a part. Access may also be affected by factors such as gender, caste, race and occupation. Small-scale producers, vulnerable and marginalized groups and women may find securing access particularly challenging.

Thus, as ample empirical evidence points out, there is a great need for nuanced, culturally and contextually appropriate and inclusive water tenure to enable sustainability (Brisbois and de Loë 2016). Any plans must factor in relevant infrastructure and access to relevant sustainable techniques, from rainwater harvesting and drip irrigation to catchment systems, agroforestry and terracing (see, e.g., Giordano et al. 2017).

As with access to other parts of nature, secure access to water can form a virtuous circle with agroecological transitions. The right to water is necessary for food and nutritional security, in which agroecology plays a major role; in turn, agroecological practices can boost water security by improving the way water is conserved and used, thus increasing resilience to water stress (HLPE 2019; Kremen and Miles 2012). For farmers, pastoralists, fishers and indigenous peoples, the ability to experiment with and implement water-conserving agroecological methods ultimately leads to more secure, diversified and resilient local livelihoods.

Seeds and Biodiversity

Agroecology focuses on indigenous, locally adapted, genetically diverse and traditional crops, so access to such seeds is central to the practice. Rights to such seeds, and decentralized, community-led seed saving, selection and plant breeding, hold great potential for innovations, resilience and livelihoods in agroecology (Halpert and Chappell 2017; Mulvaney 2020). Published research on participatory management of livestock diversity and breeding is limited (Conroy 2008), but considering their promise and the amount of relevant unmet needs in this area, models such as these also need significantly increased resources in terms of supportive policy and monetary and human resources for careful research and *implementation*. Seeds have received significant attention over the years, but as millions of people depend on livestock for livelihoods and sustenance as well, participatory livestock breeding and diversity efforts have great potential to address unmet needs in breeding for ends related to sustainability, resilience, agroecology and local adaptation (Hoffmann; Wallace 2016, p. #1029). In fisheries, a substantial literature on common property management regimes has shown the power and potential of such localized and bottom-up schemes (d'Armengol et al. 2018).

To ensure rights to seeds and biodiversity, enterprises, networks and exchanges linked to smallholder seed and livestock breeding must be protected from dominant, 'Western' intellectual property regimes, such as the typical approach taken to patents in the United States and Europe. These regimes are inappropriate in many cases where communities have cooperatively managed their own 'intellectual property' for decades or centuries without Western property rights and where requirements like variety's novelty or stability block existing, but effective, approaches and varieties from protection. Along with intellectual property rights (IPR)—the rules and rights that determine who nominally owns genetic material—the legal frameworks that dictate what seeds and livestock breeds can be sold on the market are also of paramount importance. Such globalized, Western-developed and business-focused 'protection' for seeds, germplasm and animal breeds are unsuited to agroecology and need to change in favour of local intellectual property systems (Halpert and Chappell 2017; Forsyth 2013)

Miranda Forsyth (2016) notes that customary systems for protecting intellectual property should not be based on "romantic visions of returning to some imaginary pre-colonial past, but rather on the pragmatic reality that these existing systems are already culturally attuned to promoting goals such as knowledge diffusion and promotion of creativity and innovation". To illustrate this point, she discusses a variety of customary institutions in the Pacific Islands, such as *tabu* (secrecy regimes), *talanoa* (oral research and data exchanges), *tufuga* (traditional crafts guilds) and other biocultural protocols that preserve and protect not just traditional varieties but also traditions and customs, including those governing interactions between humans and nature.

A number of traditional and new concepts and institutions hold potential for protecting access to agricultural biodiversity (whether in livestock, crops, trees or fish) and so enabling agroecological transformation. These include seed sovereignty and the Open Source Seed Initiative (Montenegro de Wit 2017), as well as Livestock Keepers' Rights, fisheries co-management (the joint management of resources by direct users, governments and other actors; see d'Armengol et al. 2018), commonification/decommodification (moving towards commons-based models and away from commodification of agriculture and its products; see Vivero-Pol et al. 2018) and the processes and proposals of the VGGT reviewed above. In general, agroecology demands approaches that not only help sustain the nominal or numerical diversity of crop varieties and livestock but also conserve and enhance high levels of genetic heterogeneity *within* each crop variety and animal breed. This is because agroecology is based on adaptive responses to diverse 'mosaics' of conditions, including complex, novel and changing ecosystems in specific places, using available genetically heterogeneous plants and animals. As we have seen, this is a circular process. That is, a diversity of agroecological practices can in turn enhance biodiversity (Mulvaney 2020).

Another key element in this context is the synergy that may develop between farm and wild biodiversity. Research has shown that greater biodiversity on a farm correlates with greater biodiversity in the surrounding environment, and that greater overall (on-farm and wild) biodiversity benefits agriculture (Palomo-Campesino et al. 2018)—for instance, in the thoroughness and efficiency of pollination, leading to greater production and better quality produce, mitigation of and greater resilience to extreme weather events (e.g. rather than attempting to breed a single variety resistant to both drought and waterlogging, a diverse system with varieties that are able to deal with many different circumstances), larger populations and greater diversity of natural predators, healthier soils and decreased erosion. Such holistic systems, with their intersecting biodiversity, have many characteristics of commons (see Box 4.1).

Politically, the approaches to accessing seeds and biodiversity outlined here can be powerful tools for supporting and enabling agroecology. More broadly, localized, participatory 'commons' regimes offer powerful examples too (see Box 4.1), which are particularly well suited to community self-organization for agroecological transformation.

Disabling Conditions

A number of disabling conditions, or 'lock-ins', pose challenges to those seeking access to ecosystems in the context of agroecological transformations.

Land

Insecure land tenure and the fragmentation of land through purchase by outside, often corporate, concerns, division during inheritance, expropriation for logging, mining, and other uses, all pose major obstacles to progress on agroecology in many countries. In Uganda, for example, farmers vulnerable to eviction are unlikely to invest in agroecology, or other activities with long-term payoffs, because of the risk of losing access to the land after making the investment (Isgren 2016). And in Bangladesh, while the government has promoted homestead (subsistence) gardening, the fragmentation of residential plots has left many rural people with a plot just big enough for the house. Unable to produce their own food, they depend extensively on the market, where nutritious foods are often unaffordable (Misra 2017).

Inequality in land tenure is pervasive globally. As measured by the Gini coefficient of inequality, where 0 is perfect equality and 1 is perfect inequality, median land ownership in Latin American countries measured was approximately 0.8; only East and Southeastern Asia and sub-Saharan Africa had Gini coefficients of less than 0.5 (Frankema 2005). Western and Northern Africa, the Caribbean, Latin America and some countries in Oceania all had coefficients close to or above 0.65, the level at which Olivier De Schutter (2010) argues that land reform may be necessary to contribute to the right to food, particularly where landlessness and small plot sizes are linked to significant levels of rural poverty.

These challenges, if combined with undue focus on export-driven policies, can disable attempts to gain land tenure for agroecology. All of these challenges arise "primarily from a dominant model of agricultural development that rewards the most mechanized and capital-intensive farms" and that generally favours large producers (De Schutter 2010), further contributing to the lock-in of the dominant agricultural regime. In urban and peri-urban areas as well, difficulties in accessing or affording land are a significant constraint on urban agroecology's potential (Tornaghi and Dehaene 2020).

Vested interests such as large-scale, wealthy land owners within many countries and international corporate, agricultural and government concerns continue to block reforms such as land redistribution, which have been scaled back or stopped in the face of strong external pressure from the effects of neoliberal capitalism or what has been called "accumulation by dispossession" (Gebara 2018; see also Borras 2007). Corporations and many nation-states see present arrangements as being to their advantage. The private property model concentrates land ownership "in the hands of fewer and fewer people, usually men", and looks to "rule and order" rather than need (Courville 2006). The current economic system (including elite opposition to land reform) is arguably working exactly as it was designed to work—which is not to the benefit of agroecological transition, smallholder farmers or the environment (Holt-Giménez 2017).

Linked to, but extending beyond, land reform and tenure is the issue of large-scale land acquisitions, also known as land grabbing, where corporations or states dispossess and often violently displace peasants from their farms and common lands (Box 4.2). As Matias Margulis et al. (2013) note:

> This global land rush is characterized by transnational and domestic corporate investors, governments, and local elites taking control over large quantities of land (and its minerals and water) to produce food, feed, biofuel, and

Box 4.2 Agroecology as an Organizing Force Against Land and Water Grabbing

Large-scale acquisition of resources, particularly land and groundwater grabbing, is threatening farmer societies in Senegal. It sweeps away all the gains achieved among rural communities who are working hard to advance agroecology. Over the past decade, these Senegalese organizations have developed a joint vision for viable and sustainable production systems and community-led land governance. In 2010 the government included the concept of 'healthy and sustainable agriculture' in its agricultural policy and earmarked a specific budget for it.

However, progress was disrupted when the same government pushed for the establishment of multinational corporations in these territories, arguing that it was the only way towards food security. This transformed family farmers into farm workers on their own land while putting the environment at risk. While only 6 cases of land grabbing (totalling 168,964 hectares) were recorded in Senegal between 2000 and 2007, there were 30 cases recorded between 2008 and 2011, accounting for a staggering 630,122 hectares. This is an unprecedented increase, and it sparked outrage and led to protests.

Existing agro-industrial facilities and mining companies have often failed to carry out environmental studies, particularly with respect to the contamination of water with chemicals and other effects on water resources. Depletion of various layers of groundwater has occurred as a result of excessive water extraction by agribusinesses. Early signs of conflict over water have emerged precisely in the areas where agroecology has taken root: Niayes, Keur Moussa, the lower valley/Lac de Guiers and the Petite Cote.

For years, farmers' organizations and their allies in Senegal have combated the co-optation of their resources through awareness raising, calls for mobilization, research, training and advocacy. The basic principle defended by farmers is that land and other resources must be in the hands of the communities and that an agricultural policy must be based on a system of financing that is favourable to family farming. Farmers' organizations developed their own policy proposals, pointing out that land must be considered along with access to nature more broadly. They called on the government to implement

(continued)

> **Box 4.1   (continued)**
>
> an integrated rural development policy to achieve food sovereignty. As a result of this advocacy work, the National Commission for Land Reform (CNRF) decided to integrate civil society organizations into its steering committee and to stop promoting the commercialization of land.
>
> These technical, organizational and political results encourage farmers' groups to pursue their mission of supporting rural families in reclaiming governance of their land and the implementation of integrated development strategies that can lead them towards food sovereignty. The situation in Senegal demonstrates that agroecology can be a strong organizational force for the protection of access to nature.
>
> *Source*: Brun (2018)

other industrial commodities for the international or domestic markets. Such land deals are often associated with very low levels of transparency, consultation, and respect for the rights of local communities living off the land.

Often taking place under the guise of investments for development, land and water grabs are arguably the opposite of much-needed land reforms for agroecology.

Water

Inevitably, issues with control of water overlap with those related to control of land. Excessive control by private interests and denial of access to local communities (Swyngedouw 2005; HLPE 2015) block the possibilities offered by agroecology. As Howard Mann and Carin Smaller (2010) wrote, "The current land purchase and lease arrangements are largely about shifting land and water uses from local farming to essentially long-distance farming to meet home state food and energy needs."

Such a switch militates against the benefits and ethos of agroecology, which emphasize strengthening local governance and local production systems. In 2014, social movements presented a declaration, Rights to Water and Land, a Common Struggle (World Forum for Alternatives 2014) [https://commonstransition.org/rights-to-water-and-land-a-common-struggle/>], at the World Social Forum in Tunis. This recognizes the "essential linkage between [their] struggles, given the inextricable nature of land and water grabbing" and states that "the scarcity underpinning the water, land and food crises is not a given; it is a political, geostrategic and financial construct".

Seeds and Biodiversity

Big industrial agriculture and the prevailing food system (the corporate food regime) depend on industrial inputs and processing. For over two decades, the markets for these products and processes have become increasingly concentrated and consolidated, a process that locks industrial methods into agriculture and disables access to seeds and biodiversity.

For example, the commercial seed market is now nearly dominated by just two firms (Howard 2016). High degrees of concentration and consolidation are also seen in markets for livestock, particularly 'broiling' chickens, turkeys, pigs and beef cattle (Hendrickson et al. 2017). Diversity within commercial livestock appears to have suffered as a result, with "only two companies providing layer hen genetics and four providing those for broilers": significant volumes of global egg and broiler production are thus now designed to meet industrial needs (Gura 2007). This trend has been called "one of the most pressing concerns" about the industrialization of agriculture (Hendrickson et al. 2017). Narrowing diversity, both within and among seeds and breeds, is hugely problematic for agroecologists, as corporate concerns extend their grasp around the world, and traditional seed- and breed-saving methods and networks are marginalized or even rendered illegal.

We discussed earlier international intellectual property regimes based primarily on legalistic Western approaches. These are arguably ill-suited to help farmers adapt to climate change and extreme weather events, particularly as they tend towards doubling down on failed strategies by seeking to produce monocultures of drought-resistant seeds—seeds that then do not have the genetic variation or the legal entitlements necessary for them to adapt and respond to other foreseen or unforeseen challenges, from waterlogging to novel pests (Halpert and Chappell 2017). The control of so much of the seed supply by a handful of international corporations stifles farmers' innovation by centralizing research and drastically restricting farmer experimentation and variety development through intellectual property 'protections'. This limits their ability to adapt seeds to local conditions by locking useful traits and varieties behind intellectual property, buying up or undercutting local seed systems, and promoting 'faddism'—or in the words of some analysts, draining more from the pool of knowledge (through patents) than they are giving back (Halpert and Chappell 2017; Stone et al. 2014). Similar problems arise in livestock rearing (Wallace 2016; Gura 2007). Taken together, these trends constitute a threat to the practice of viable agroecology.

References

Araghi, F. (2008). The Invisible Hand and the Visible Foot: Peasants, Dispossession and Globalization. In A. H. A.-L. C. Kay (Ed.), *Peasants and Globalisation: Political Economy, Rural Transformation and the Agrarian Question* (pp. 111–147). London: Routledge.

Borras, S. M., Jr. (2007). *Pro-poor Land Reform: A Critique*. Ottawa: University of Ottawa Press.

Brisbois, M. C., & de Loë, R. C. (2016). Power in Collaborative Approaches to Governance for Water: A Systematic Review. *Society & Natural Resources, 29*(7), 775–790.

Brun, L. (2018). Landgrabbing Threatens Agroecology in Senegal. *Farming Matters.*

Chappell, M. J., Wittman, H., Bacon, C. M., Ferguson, B. G., Barrios, L. G., Barrios, R. G., et al. (2013). Food Sovereignty: An Alternative Paradigm for Poverty Reduction and Biodiversity Conservation in Latin America. *F1000Res, 2*, 235.

Civil Society Mechanism for Relations to the Committee on World Food Security (CSM). (2016). *Synthesis Report on Civil Society Experiences Regarding Use and Implementation of the Tenure Guidelines and the Challenge of Monitoring CFS Decisions.* Rome.

Collins, A. M. (2014). Governing the Global Land Grab: What Role for Gender in the Voluntary Guidelines and the Principles for Responsible Investment? *Globalizations, 11*(2), 189–203.

Conroy, C. (2008). Livestock, Livelihoods, and Innovation. In S. S. S. B. Pound (Ed.), *Agricultural Systems: Agroecology and Rural Innovation for Development* (pp. 253–280). Burlington, MA: Elsevier.

Courville, M. P. R. (2006). Introduction and Overview: The Resurgence of Agrarian Reform in the Twenty-first Century. In R. C. P. M. C. P. M. Rosset (Ed.), *Promised Land: Competing Visions of Agrarian Reform* (pp. 3–22). New York: Food First Books/CDS.

d'Armengol, L., Castillo, M. P., Ruiz-Mallén, I., & Corbera, E. (2018). A Systematic Review of Co-managed Small-Scale Fisheries: Social Diversity and Adaptive Management Improve Outcomes. *Global Environmental Change, 52,* 212–225.

De Schutter, O. (2010). *The Right to Food: Interim Report of the Special Rapporteur.* New York, NY: The United Nations.

Deininger, K., Jin, S., & Nagarajan, H. K. (2009). Land Reforms, Poverty Reduction, and Economic Growth: Evidence from India. *The Journal of Development Studies, 45*(4), 496–521.

Duncan, J. (2015). *Global Food Security Governance: Civil Society Engagement in the Reformed Committee on World Food Security.* Oxon: Routledge.

Forsyth, M. (2016). Making the Case for a Pluralistic Approach to Intellectual Property Regulation in Developing Countries. *Queen Mary Journal of Intellectual Property, 6*(1), 3–26.

Forsyth, M. F. S. (2013). Intellectual Property and Food Security in Least Developed Countries. *Third World Quarterly, 34*(3), 516–533.

Frankema, E. H. (2005). *The Colonial Origins of Inequality: Exploring the Causes and Consequences of Land Distribution.* Discussion papers//Ibero America Institute for Economic Research.

Fraser, E. D. G. (2004). Land Tenure and Agricultural Management: Soil Conservation on Rented and Owned Fields in Southwest British Columbia. *Agriculture and Human Values, 21*(1), 73–79.

Gebara, M. F. (2018). Tenure Reforms in Indigenous Lands: Decentralized Forest Management or Illegalism? *Current Opinion in Environmental Sustainability, 32,* 60–67.

Giordano, M., Turral, H., Scheierling, S. M., Tréguer, D. O., & McCornick, P. G. (2017). *Beyond "More Crop per Drop": Evolving Thinking on Agricultural Water Productivity.* Colombo, Sri Lanka: IWMI Research Report No. 169.

Gura, S. (2007). *Livestock Genetics Companies: Concentration and Proprietary Strategies of an Emerging Power in the Global Food Economy.* Ober-Ramstadt, Germany: League for Pastoral Peoples and Endogenous Livestock Development.

Halpert, M.-T., & Chappell, M. J. (2017). *Prima facie* reasons to question enclosed intellectual property regimes and favor open-source regimes for germplasm [version 1; peer review: 3 approved, 1 approved with reservations]. *F1000Research, 6*:284. https://doi.org/10.12688/f1000research.10497.1.

Hardin, G. (1968). The Tragedy of the Commons. *Science, 162*(3859), 1243–1248.

Hendrickson, M. K., Howard, P. H., & Constance, D. H. (2017). Power, Food and Agriculture: Implications for Farmers, Consumers and Communities. Division of Applied Social Sciences Working Papers, University of Missouri College of Agriculture, Food & Natural Resources.

Heredia, B., Medeiros, L., Palmeira, M., Cintrão, R., & Pereira Leite, S. (2016). Regional Impacts of Land Reform in Brazil. In R. C. P. P. M. Rosset & M. Courville (Eds.), *Promised Land: Competing Visions of Agrarian Reform* (pp. 277–300). New York: Food First Books/CDS.

Higgins, D., Balint, T., Liversage, H., & Winters, P. (2018). Investigating the Impacts of Increased Rural Land Tenure Security: A Systematic Review of the Evidence. *Journal of Rural Studies, 61*, 34–62.

HLPE. (2015). *Water for Food Security and Nutrition: A Report by the High Level Panel of Experts on Food Security and Nutrition of the Committee on World Food Security.* Rome: HLPE.

HLPE. (2019). *Agroecological and Other Innovative Approaches for Sustainable Agriculture and Food Systems That Enhance Food Security and Nutrition.* Rome: High Level Panel of Experts on Food Security and Nutrition of the Committee on World Food Security.

Holt-Giménez, E. (2017). *A Foodie's Guide to Capitalism: Understanding the Political Economy of What We Eat.* New York: Monthly Review Press and Food First Books.

Howard, P. H. (2016). *Concentration and Power in the Food System: Who Controls What We Eat?* London: Bloomsbury Academic Publishing.

Isgren, E. (2016). No Quick Fixes: Four Interacting Constraints to Advancing Agroecology in Uganda. *International Journal of Agricultural Sustainability, 14*(4), 428–447.

Kremen, C., & Miles, A. (2012). Ecosystem Services in Biologically Diversified Versus Conventional Farming Systems: Benefits, Externalities, and Trade-offs. *Ecology and Society, 17*(4), 40.

Lawry, S., Samii, C., Hall, R., Leopold, A., Hornby, D., & Mtero, F. (2016). The Impact of Land Property Rights Interventions on Investment and Agricultural Productivity in Developing Countries: A Systematic Review. *Journal of Development Effectiveness, 9*(1), 61–81.

Lipton, M. (2009). *Land Reform in Developing Countries: Property Rights and Property Wrongs.* London and New York: Routledge.

Lipton, M., Yaqub, S., & Darbellay, E. (1998). *Successes in Anti-poverty*. Geneva: International Labor Organization.

Mann, H., & Smaller, C. (2010). Foreign Land Purchases for Agriculture: What Impact on Sustainable Development. *Sustainable Development Innovation Brief, 8*, 1–8.

Margulis, M. E., McKeon, N., & Borras, S. M. (2013). Land Grabbing and Global Governance: Critical Perspectives. *Globalizations, 10*(1), 1–23.

Misra, M. (2017). Moving Away from Technocratic Framing: Agroecology and Food Sovereignty as Possible Alternatives to Alleviate Rural Malnutrition in Bangladesh. *Agriculture and Human Values, 35*, 473–487.

Montenegro de Wit, M. (2017). Beating the Bounds: How Does 'open source' Become a Seed Commons? *The Journal of Peasant Studies, 46*(1), 44–79.

Mulvaney, P. (2020). Sustaining Agricultural Biodiversity and Heterogeneous Seeds for Responsible Agriculture and Food Systems. In A. Kassam & L. Kassam (Eds.), *Rethinking Food and Agriculture*. Elsevier.

Ostrom, E. (1990). *Governing the Commons: The Evolution of Institutions for Collective Action*. Cambridge University Press.

Otsuka, K. P. F. M. (2001). *Land Tenure and Natural Resource Management: A Comparative Study of Agrarian Communities in Asia and Africa*. International Food Policy Research Institute (IFPRI): Baltimore and London.

Palomo-Campesino, S., González, J., & García-Llorente, M. (2018). Exploring the Connections between Agroecological Practices and Ecosystem Services: A Systematic Literature Review. *Sustainability, 10*(12), 4339.

Ribot, J. (2014). Cause and Response: Vulnerability and Climate in the Anthropocene. *The Journal of Peasant Studies, 41*(5), 667–705.

Robbins, P. (2011). *Political Ecology: A Critical Introduction* (Vol. 16). John Wiley & Sons.

Schwendler, S. F., & Thompson, L. A. (2017). An Education in Gender and Agroecology in Brazil's Landless Rural Workers' Movement. *Gender and Education, 29*(1), 100–114.

Stone, G. D., Flachs, A., & Diepenbrock, C. (2014). Rhythms of the Herd: Long Term Dynamics in seed choice by Indian farmers. *Technology in Society, 36*, 26–38.

Swyngedouw, E. (2005). Dispossessing H2O: The Contested Terrain of Water Privatization. *Capitalism Nature Socialism, 16*(1), 81–98.

Tornaghi, C., & Dehaene, M. (2020). The Prefigurative Power of Urban Political Agroecology: Rethinking the Urbanisms of Agroecological Transitions for Food System Transformation. *Agroecology and Sustainable Food Systems, 44*:5, 594–610, https://doi.org/10.1080/21683565.2019.1680593

Vivero-Pol, J. L., Ferrando, T., De Schutter, O., & Mattei, U. (2018). *Routledge Handbook of Food as a Commons*: Routledge.

Wallace, R. (2016). *Big Farms Make Big Flu: Dispatches on Influenza, Agribusiness, and the Nature of Science*. New York, NY: Monthly Review Press.

Wittman, H. (2010). Agrarian Reform and the Environment: Fostering Ecological Citizenship in Mato Grosso, Brazil. *Canadian Journal of Development Studies/ Revue canadienne d'études du développement, 29*(3–4), 281–298.

Woodley, E., Crowley, E., de Pryck, J. D., & Carmen, A. (2006). *Cultural Indicators of Indigenous Peoples' Food and Agro-Ecological Systems.* SARD Initiative Commissioned by FAO and the International India Treaty Council.

World Forum for Alternatives. (2014). Rights to Water and Land, a Common Struggle.

CHAPTER 5

Domain B: Knowledge and Culture

Abstract In this chapter, we examine the role of knowledge processes in the form of local practice, research, innovation and education in agroecology transformations. Knowledge and power are intimately linked; the questions of 'what knowledge' and 'whose knowledge' is valued are vitally important. We review the informal (outside of institutions) and formal knowledge processes that have been found to support agroecology. These affirm and enable the knowledge systems of agricultural producers, especially those of women and youth. We further discuss how the combination of scientific knowledge with local and traditional knowledge is important in agroecology transformations. Unfortunately, mainstream knowledge systems often disable agroecology because they privilege outside and top-down processes of knowledge transfer that invalidate local, farmer and indigenous knowledges.

Keywords Research • Learning • Education • Cognitive justice • Peasant-to-peasant

The way knowledge is constructed, produced, shared and put to use is critically important in any shift to agroecology (Levidow et al. 2014; IAASTD 2008). But the knowledge required for agroecology is radically different from that available in mainstream institutions such as agricultural universities and policy think tanks (Pimbert 2018). This is mostly because agroecology develops in very context-specific ways. Knowledge of the local ecosystem, history and landscape is crucial, and much of that is in the hands of farmers and other local people, as well as often deeply linked with cultural practices. Agroecological transformation emerges from the interactions between cultural identity and the knowledge inherent in culturally grounded practice.

Thus, knowledge and power are intimately linked. What knowledges are enabled and valued? Who are acknowledged as valid holders and producers of knowledge? The answers are crucial in shaping the potential of transformations in agroecology. Struggles over knowledge play out across related areas of research, innovation and education. Here, the cosmovisions, epistemologies and validity of alternative agroecological knowledge systems come into conflict with the scientist and corporate control of the dominant knowledge system. In this context, the remit of learning and knowledge goes far beyond the adoption of specific agroecological techniques. Instead, it depends on, and in turn reinforces, wider processes of democratization, organization and inclusion. Agroecological knowledge, like all of these domains, is therefore deeply entwined with issues of governance.

ENABLING CONDITIONS

A great deal of agroecological knowledge, learning and innovation is produced, held and mobilized (deployed in political process of transformation) in the networks and organizations of indigenous peoples and food producers (see Boxes 5.1 and 5.2). This points to a need to actively transform and construct knowledge systems to reflect diversity and decentralization, promote dynamic adaptation and deepen democracy. Traditional ecological knowledge, indigenous knowledge and the knowledge of agroecological farmers must be brought into dialogue with scientific ways of knowing.

Such a shift demands a departure from the linear 'knowledge transfer' approach dominated by formal science and experts (Pimbert 2018). Agroecological knowledge must be developed through complex and ongoing processes centred on social learning through networks of diverse actors engaged in knowledge dialogues.

Traditional Knowledge and Culture

As we have seen above, agroecology is highly context specific; to be effective, it must be based on place-based, lived knowledge. Thus, traditional ecological knowledge (TEK) is central within it. TEK is in essence knowledge held by a society or culture that is related to their local environment. Globally, many traditional agricultural systems based on TEK resonate with agroecological principles, from East Asian rice-fish systems (combining aquaculture and rice cultivation) to Mexico's *milpa-solar* cropping systems (cultivating maize, beans and squash together on home plots called *solares*) or Andean *waru-waru* ridge fields, which control drought and frost.

The indigenous knowledge imbuing these agroecological systems is deeply intertwined with their cultural practices associated with managing and protecting forests and other ecosystems for wild food and medicinal-plant gathering (Woodley et al. 2006). Many indigenous cultures pass traditional knowledge and genetic resources from generation to generation through ceremonies, stories, songs and oral histories.

Unfortunately, these practices and the knowledge associated with them are often viewed as antiquated, with little value for modern agriculture. Centuries of colonialism have also eroded them, not least through Western bias in development and research and the imposition of corporate knowledge and technologies. To enable agroecology, conserving and reviving traditional stores of local knowledge and practice are critical. So are enabling the cultural practices associated with stewarding biodiversity and territories and identifying the factors that impede or encourage indigenous elders' transferal of knowledge to younger generations (Woodley et al. 2006, p. 11).

The Globally Important Agricultural Heritage Systems (GIAHS) initiative, started in 2002 under the auspices of the Food and Agriculture Organization of the United Nations (FAO), now encompasses 59 sites in 22 countries that are recognized for their function as reservoirs of biodiversity, culture and traditional knowledge. The GIAHS initiative reflects an important intergovernmental commitment to preserving these knowledge systems. However, beneficial traditional practices also exist outside these recognized landscapes and it is important to acknowledge, protect and harness these processes where they are present.

Horizontal Learning

Horizontal learning is based on principles of Freirian pedagogy and is based on reciprocal learning dialogues and exchanges where the hierarchy between teacher and learner is intentionally dissolved and where all actors, offering their own experience and knowledge in the learning environment, are regarded as teacher-learners. Through a horizontal approach, these actors build capacity in terms of agroecological practice and politics but also as teachers, thus enabling the ongoing spread of agroecology in a horizontal pattern.

Colin Anderson et al. (2019) proposed that food producers and social movements must lead any transformative agroecology learning approach and that it must be based on four key characteristics or qualities (Fig. 5.1):

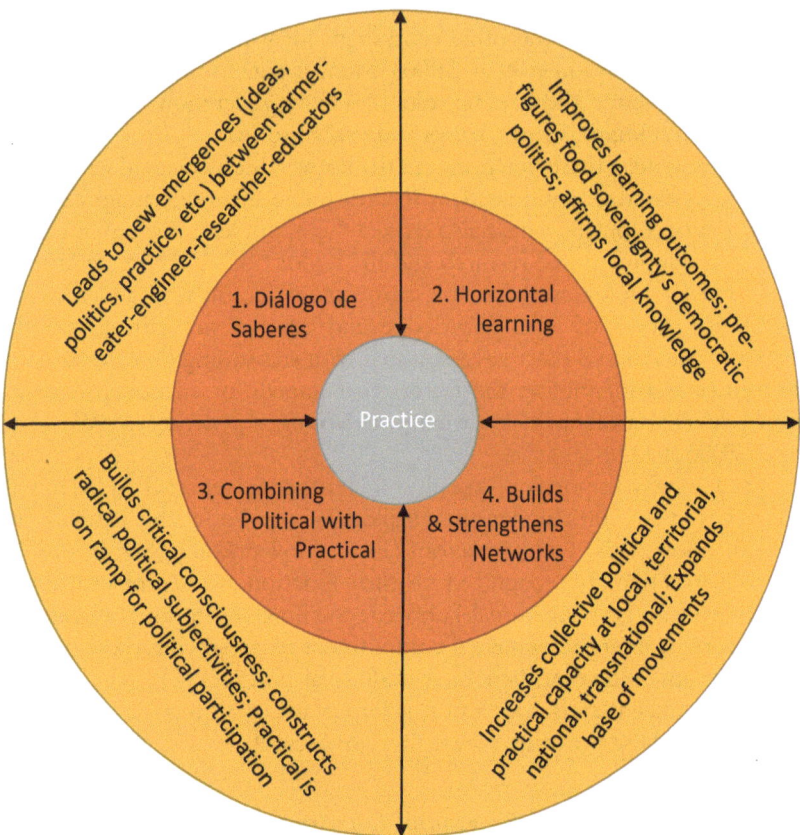

Fig. 5.1 Transformative agroecology framework by Anderson et al. (2019) involves a pedagogical approach that places practice as a central component of all training. It however integrates four pillars (the orange segments) to provide the 'connective tissue' to the political project of food sovereignty (the yellow circle)

horizontal learning; *diálogo de saberes* (wisdom dialogue); combined practical and political knowledge; and building networks. Thus, while the protagonism of food producers and farmer organizations in agroecological learning is important, it is critical that methodologies and pedagogies go beyond 'the practical' and are linked to political work underway in territories.

Horizontal forms of adult learning among agroecological producers have been found to be vital in spreading agroecology (Mier y Terán Giménez Cacho et al. 2018; Anderson et al. 2019; McCune and Sánchez 2018). A well-known example is the farmer-to-farmer (campesino a campesino or CaC, Box 5.1) methodology that originated in the 1980s in Central America. Under it, farmers come together to discuss their farms, lives and dilemmas and diagnose and solve problems collectively. Peter Rosset (2011, P. 169), discussing the CaC approach in Cuba, explains the effectiveness of the farmer-driven social process of CaC:

> A fundamental tenet of CaC is that farmers are more likely to believe and emulate a fellow farmer who is successfully using a given alternative on their own farm than they are to take the word of an agronomist of possibly urban extraction. Whereas conventional extension can be demobilizing for farmers, CaC is mobilizing, as they become the protagonists in the process of generating and sharing technologies.

Box 5.1   Movimiento Campesino a Campesino: Practical and Political Learning from Farmer to Farmer

The Movimiento Campesino a Campesino (CaC), or Farmer-to-Farmer Movement, is one of the earliest and most successful efforts for promoting sustainable agriculture in Latin America. The extensive knowledge networks that are at the basis of the CaC methodology have been highly effective not only in generating and spreading sustainable agricultural practices on the ground but also in enabling farmers to build skills and organizational capacity.

CaC involves hundreds of volunteer and part-time campesino *promotores* working with thousands of farmers and with the support of dozens of technicians, professionals and local development organizations. They have used relatively simple small-scale experimentation, combined with horizontal (farmer-to-farmer) workshops in basic ecology, agro-pastoralism, agronomy, soil and water conservation,

(continued)

Box 5.1 (continued)

Fig. 5.2 Campesino a campesino learning in Latin America

soil building, small-scale livestock care, seed selection, crop diversification, integrated pest management and biological weed control (including integrating livestock). These approaches provide farmers with sufficient technical and ecological knowledge, as well as the necessary conviction, enthusiasm and pride to reverse degenerative processes and overcome the basic limiting factors in farm production. CaC has succeeded in regenerating tens of thousands of hectares of exhausted soils in the tropics.

Over the years, new insights emerged about the urgent necessity to accompany such practical farmer-to-farmer learning with structural social and political change. In an effort for agroecological experiences to spread both geographically and into the institutions that structurally shape agriculture's social, economic and political terrain, more recent CaC initiatives have included political training and

(*continued*)

> **Box 5.1 (continued)**
>
> organizing. The aim is to confront the structures and policies pre-
> venting the spread of agroecology and influence them to support it
> instead.
>
> CaC has proven to be an effective social-organizational approach
> to developing a 'cadre' of agents capable of working collectively in
> technical and political work—such as social movement base-building,
> education or lobbying—to encourage the spread of agroecology.
> Building on these insights, the global peasant movement La Via
> Campesina (LVC) is creating learning processes centring on experi-
> mentation, innovation, recollection, sharing and the spread of agro-
> ecological methods under the umbrella of 'agroecology schools'. All
> of its schools combine technical and political education, as well as
> practice and theory. Their connection of CaC processes with schools
> for permanent training and practice-based reflection is a strong strat-
> egy for scaling agroecology out. LVC currently operates some 65
> such schools around the world.
>
> *Source:* La Via Campesina (2017); Holt-Giménez (2006)

Participatory and popular educational models of collective learning
such as CaC are critical for the development of agroecology and for decen-
tralizing agricultural development. Thus, building on the work of popular
educators such as Paulo Freire, horizontal learning approaches by grassroots organizations in Brazil, Cuba, Nicaragua, India and other countries
become self-perpetuating: they spread organically, as learners become
teachers and farmers' organizations and social movements become stronger. The methods encourage deeper reflexive learning and in recent years
have started to be linked with political training (McCune and Sánchez
2018), increasing the collective capacity of the networks in agroecology to
influence the dominant food system. When technical education in agroecological practice is embedded in and linked to organizing within social
movements, participants can develop a collective self-awareness of their
situations, link lessons learned in local projects and articulate joint aspirations and demands.

Farmer Field Schools (FFSs), pastoralist field schools and other such bodies have been championed by FAO and taken up by NGOs and farmer groups around the world. These too have been a powerful way of spreading and deepening knowledge of, for and by agricultural producers. The schools bring together groups of agricultural producers to develop appropriate solutions based on methods such as agro-pastoralism, conservation agriculture, organic farming and integrated pest management (IPM). In over 90 countries, FFSs have enabled agricultural producers to build knowledge, reduce pesticide and other external input use and shift towards more sustainable livelihoods. The field school approach emphasizes empowerment through increasing capacity among farmers and is based on farmer-centred experiential learning in the field. This allows producers to collectively observe, measure, analyse, assess and interpret agroecosystem relationships in decision-making.

In Indonesia, FFSs were effective in the spread of IPM. Their emphasis on marrying knowledge-work (training and research) and the strengthening of farmer's organizations and networks was critical in farmers' ownership of IPM across the country while simultaneously supporting the peasant movement. Interestingly, over time, a programme initially promoted by the field school initiated by FAO and the government evolved into what was then called 'community IPM'. This shifted the locus of agency to communities, emphasizing collective organization and institutionalizing IPM locally through horizontal FFS networks (Fakih et al. 2003). This, too, highlights the importance of embedding learning and training within broader networks of social movements and farmers' organizations (Anderson et al. 2019; see also Chap. 7).

Box 5.2   Agroecology Training Programme of the Coordination Nationale des Organizations Paysannes (CNOP) in Mali
The Coordination Nationale des Organisations Paysannes (CNOP) is a federation of 13 national peasant farmers' organizations in Mali.[1] In 2011 the CNOP launched a training programme for farmer-trainers to scale out agroecology by building Malian

(continued)

Box 5.2   (continued)

Fig. 5.3 Malian farmers at the Nyéléni International Center for Training in Peasant Agroecology (*Photo credit*: Colin Anderson)

farmers' capacity in peasant agroecology. The farmer-trainers are producers recognized by their peers. First, they participate in the training at the Nyéléni International Center for Training in Peasant Agroecology, located 140 km from Bamako. Then they share this knowledge in their cooperative, village or local producers' group.

Training sessions bring together farmers, pastoralists and fisherfolk from different regions and age groups, with an equal number of women and men, to share information on agroecological practices. Trainings are interactive, fostering the exchange of peasant knowledge, know-how and way of life. They cover local and global issues, struggles to be waged and agricultural policies for embedding territories in peasant agroecology.

These mechanisms—mixing participants from various agricultural sectors and different geographical areas and the equal number of male/female participants with at least 40% of young people—help to

(*continued*)

Box 5.2 (continued)

convey common messages and build knowledge collectively through horizontal forms of learning. By valuing each person's knowledge and skills, these educational choices help reduce inequality as well as social and cultural pressures while strengthening self-confidence and equity among participants.

Today, there are more than 500 farmer-trainers in peasant agroecology who have trained 20,000 producers across Mali. Trainees are often at the heart of informally organized regional commissions that maintain close links with the CNOP and help spur local economic activity around production units like collective fields worked by women and processing units for local juices, soap and parboiled rice. Through the trainees, these commissions help translate the spirit of trainings into concrete outcomes, creating synergy, innovation, income, employment, autonomy and dignity. From the training, a whole territory is on the move, mobilizing peasant organizations and communities and amplifying the positive impact of peasant agroecology in Mali.

Prepared in collaboration with Chantal Jacovetti, formerly responsible for peasant agroecology and land-related issues at the CNOP, currently working as a consultant on these themes and more broadly on agriculture in Mali and West Africa.

Learning across territories is important if agroecology is to make an impact within dominant food systems, because the integration and adaptation of knowledge from other places is key to sharing innovations. The international peasant movement LVC, for instance, has developed a worldwide agroecology learning network through peasant-to-peasant processes which have been described as the "motor" of agroecological scaling (Val et al. 2019). This advances agroecological knowledge from the perspective of farmers' experiences in their own territories, then disseminates it among territories, regions and countries. LVC has become, with other social movements and food producer organizations, a key protagonist in developing agroecological knowledge and mutual learning (Box 5.1).

Diálogo de saberes

While the knowledge systems of traditional cultures, civil society and farmers are important, they are most powerful in dialogue with expert and scientific knowledge. A transformative agroecology is not anti-expert: instead, it demands that local and experiential knowledge be seen as equally important in agroecological development. In this context, the role of knowledge professionals such as researchers, teachers and technicians is oriented less towards spreading knowledge and more towards enabling the process of knowledge-sharing—especially co-producing knowledge *with* communities rather than producing and extending it *to* them. It is the job of knowledge professionals to provide advice and support to validate and improve agroecological practice, preferably through the collaborative co-production of knowledge with farmers (Pimbert 2018).

Dialogue between food producers and scientists, agricultural extensionists and educators allows agricultural producers an active, central role in testing, fine-tuning and scaling-out agroecological knowledge and practice, thereby capitalizing on their know-how and experience. Research methodologies for agroecology therefore emphasize participatory approaches to action, learning and analysis, with an emphasis on transdisciplinary ways of knowing that mobilize knowledge for social change and involve stakeholders in research (Méndez et al. 2015; Lamine 2018).

In this context it is important, however, to note two pervasive phenomena: the power imbalance between experts and non-experts and the ongoing incentives in science such as the pressure to publish with no accountability to actors outside research structures. These factors can encourage 'extractive' relationships that primarily benefit scientific partners (Levidow et al. 2014). Effective knowledge co-creation demands a substantial shift in institutional reward structures and professional norms and culture (Pimbert 2018). Charles Levkoe et al. (2019) argue that research compatible with transformative agroecology and food sovereignty should be based on three pillars: people (humanizing research relationships), power (equalizing power relations) and change (pursuing transformative rather than technocratic goals). Fortunately, in formal agroecology training and education, attention is paid increasingly to knowledge democracy, holistic agroecology and understanding agroecology as equally rooted in science, practice and politics. New pedagogical methods fostering this understanding include, for example, student collaborations in local agroecological dynamization, experiential learning and action learning.

DISABLING CONDITIONS

The agroecological knowledge we describe above is rooted in a different logic from that of the centralized, decontextualized knowledge inherent in the dominant food system. Agroecology sits within wider agricultural knowledge, science and technology (AKST) systems, which benefit from substantial public and private funding around the world. Knowledge conceived, produced and distributed in these systems reflects norms of the dominant regime and serves its interests. This poses significant challenges to agroecology, which remains "on the margins of the agricultural sciences, as it is distant from the main scientific approach as well as from the technological regime and the larger economic and political dominant trends" (Vanloqueren and Baret 2009, p. 980).

Agricultural science, extension and education tend to validate and enable the knowledge and ways of knowing that reflect the dominant food regime while invalidating and disabling alternative knowledges (Pimbert 2018). Given the power of the private sector, notably agribusiness, the factor determining validation of knowledge is often its commercial productive potential. Over the past decades, agricultural research has increasingly focused on intensive industrial production. Public and private sectors have invested heavily in crop improvement programmes focused on bio-technological methods and agricultural chemicals, despite the arguable scarcity of the kinds of 'rigorous' evidence that is often demanded by detractors to prove the value of agroecological methods (Loevinsohn et al. 2013).

All over the world, the knowledge of farmers is increasingly being marginalized in agricultural development. International standardization and globalization have contributed to migration, poverty and the loss of local knowledge among small farmers and indigenous producers (Vogl et al. 2005). And while agroecology is in the main a transdisciplinary, producer-oriented, collective approach, mainstream knowledge systems are deeply biased towards the knowledge of officially recognized experts, compartmentalized approaches to learning and top-down technology transfer (Chambers and Jiggins 1987). Traditional farming knowledge, practices and systems are displaced through the imposition of agricultural packages that are economically and materially dependent on external knowledge, technology and inputs.

Gaëtan Vanloqueren and Phillipe Baret (2009) illustrate how current science-related policy, along with the cultural and cognitive routines of scientists and institutions, limits how much mainstream innovation systems can benefit agroecology. Science-related policies are largely oriented towards growth and national competitiveness, while in many countries, public sector agricultural research and extension have been substantially scaled back. The ongoing corporate funding and control over research reinforces these trends, along with intellectual property rights that favour private sector research and development of patents, for instance, which undermine plant and livestock breeders' rights.

In this context, science and technocratic innovation that can be commercialized are prioritized, while innovations that meet social and ecological needs struggle to gain recognition and resource. Even though sustainability is a part of the discourse in agriculture, the mainstream agricultural knowledge system has become disconnected from farm and field, with its focus on the laboratory as the wellspring of new agricultural inputs and the 'rational' use of fertilizer, pesticides, herbicides and precision genetics. Reductionist approaches to knowledge development, such as those that focus on maximizing yield of single crops, are unable to account for the complex interactions in agroecological systems or their diverse, multifunctional benefits.

In this context, the knowledge and expertise of agricultural producers, citizens, small- and medium-sized enterprises, indigenous peoples and social movements are obscured and sidelined, or worse, extracted and commercialized. This can be seen in the case of biopiracy, when the traditional knowledge and genetic heritage of indigenous peoples are exploited for commercial gain, "particularly with the expansion of biotechnology for medicines, while they receive few or no material benefits and often risk resource depletion and the loss of their food sovereignty" (Woodley et al. 2006, p. 16).

The marginalization of local, non-expert and non-scientific ways of knowing (see also Chap. 9 on the discourse domain) reflects a more deeply rooted colonial view of knowledge. In it, non-Western, traditional and women's knowledge are 'othered', devalued and in some cases systematically erased (Santos 2015). Modern development has been especially blind to the knowledge and lives of indigenous peoples, women and other marginalized actors (Woodley et al. 2006). Omar Felipe Giraldo and Peter Rosset (2018) argue how ongoing approaches to agricultural development—largely led by actors in the global north and billed as 'sustainable agriculture'—continue to dehumanize and disempower communities, rendering them targets for expert knowledge and external management.

The agricultural modernization approach violates the principles of cognitive justice (Visvanathan 2005)—that is, where the invalidation of farmers as knowledge-holders ultimately enables the imposition of many of the technologies and tenets of mainstream agriculture. The new knowledge-technology packages (e.g. artificial intelligence) are, inevitably, geared to industrialized intensive agriculture, with all its attendant problems. By marginalizing alternative production systems, this process of modernization effectively overwrites traditional and indigenous knowledge systems and the lifeworlds that they sustain.

Giraldo and Rosset's (2018) analysis is similarly critical of the process of agricultural development where communities become disempowered, stripped of their agency and capacity for self-organization by the imposition of external technologies and methods. Modern development has been especially blind to, and damaging of, the knowledge and lives of indigenous peoples. The importance of cultural practices is often invisible in the sustainable development paradigm (Woodley et al. 2006). The loss of these cultural practices and indigenous languages, along with the displacement of indigenous people from traditional lands, severs the links between culture, traditional food systems and indigenous peoples' role as ecological stewards. Globalization and modernization threaten linguistic and cultural diversity, as do educational systems based in Western traditions and assimilation policies.

Beyond production, agricultural modernization has transformed the associated cognitive frameworks and cultural dynamics of food producers and communities. Decades of development led by Western science and corporate interests has depleted traditional ecological knowledge and practice. Not only have producers become materially and economically dependent on agribusiness inputs, they have also become ideologically committed to high-input, industrial-style approaches similar to the 'green revolution' of the 1960s and 1970s. In this context, local knowledge and corporate Western knowledge are often intertwined. Transformation cannot be addressed simply by incrementally advancing participatory research and development. It needs to be part of a much wider process confronting the material, cultural and spiritual legacies of colonialism (Waldmueller 2015).

Another critical global issue is the disproportionate influence of experts and institutions in the global north who shape the research, innovation and development agenda. Today, science and innovation are predominantly developed and controlled by experts and scientists in Europe and North America. Even within agroecology research, there are substantial power imbalances. Academics in industrialized countries conduct research wherever they like but do not publish in outlets based in non-industrialized countries, thus limiting the utility of the knowledge in the local context and language. Meanwhile, academics in non-industrialized countries rarely conduct research outside of their borders and publish wherever they can (Fernando Gomez et al. 2013). While knowledge gleaned abroad plays an important role in how agroecological innovation is shared, the context-specific premise of agroecology does tip the balance towards local expertise. This includes the recognition and capacity of local research and researchers.

In addition, research on diversified agricultural production systems and agroecology is severely underfunded in most parts of the world (Carlisle and Miles 2013). In their study analysing projects funded through the 2014 US Department of Agriculture Research, Extension, and Economics budget, Marcia DeLonge et al. (2016) found that the allocation to agroecology was just 0.6–1.5% of the entire budget. Most of that was earmarked for on-farm agroecology techniques and only a small portion to the socio-economic elements such as strengthening farmer organizations and developing territorial governance systems like food policy councils that are essential for upscaling the system to achieve a sustainable food system.

When serving as UN Special Rapporteur on the Right to Food, Olivier De Schutter (2010) noted the lack of public and private sector funding for agroecology research "perhaps because [its] practices cannot be rewarded by patents"—even though they need prioritizing due to their "considerable and largely untapped potential". Without adequate funding for research, promising traditional and emerging methods of agroecology cannot be supported or analysed enough for scaling up. This lack of 'evidence' and validation has led to doubts about the efficiency and credibility of agroecological alternatives.

Another limitation to the development of agroecology in the knowledge domain is its incompatibility with the indicators commonly used to evaluate and monitor progress or success in agriculture (IPES-Food 2016; Binimelis et al. 2014). Indicators on progress tend to have a narrow focus on crop and livestock productivity and cash income. The benefits of agroecology, on the other hand, rarely lead to substantial increases in the productivity of single crops (but rather increase overall productivity and reduce external inputs) and may have non-cash benefits (through subsistence, barter or other systems of economic exchange) that are erased by conventional indicators. The many social and ecosystem functions of agroecology are rarely taken into consideration.

Other progress indicators would be more consistent if applied to the economy of family agriculture. Conventional indicators such as profitability do not reveal how the management of agroecological systems generates 'added value' that includes, but goes beyond, crop diversity. Agroforestry and silvo-pastoral systems can create carbon sinks, increase agricultural biodiversity, reduce risk, produce diverse crops and livestock, preserve soil and water, maintain landscape elements and generate profits for agricultural producers. But these benefits are complex and difficult to measure, in part because they are slow, long-term processes.

Agroecology transformations demand new methods and approaches to evaluating success, including new indicators, to monitor and recognize the complex and multifunctional benefits of agroecological approaches. In this light, FAO is currently developing and testing a global analytical framework for the multidimensional assessment of the performance of agroecology: the Tool for Agroecology Performance Evaluation.

References

Anderson, C. R., Maughan, C., & Pimbert, M. P. (2019). Transformative Agroecology Learning in Europe: Building Consciousness, Skills and Collective Capacity for Food Sovereignty. *Agriculture and Human Values, 36*(3), 531–547.

Binimelis, R., Rivera Ferre, M. G., Tendero, G., Badal, M., Heras, M., Gamboa, G., et al. (2014). Adapting Established Instruments to Build Useful Food Sovereignty Indicators. *Development Studies Research, 1*(1), 324–339.

Carlisle, L., & Miles, A. (2013). Closing the Knowledge Gap: How the USDA Could Tap the Potential of Biologically Diversified Farming Systems. *Journal of Agriculture, Food Systems, and Community Development, 3*(4), 219–225–219–225.

Chambers, R., & Jiggins, J. (1987). Agricultural Research for Resource-Poor Farmers Part I: Transfer-of-Technology and Farming Systems Research. *Agricultural Administration and Extension, 27*(1), 35–52.

Charles Z. Levkoe, Josh Brem-Wilson & Colin R. Anderson. (2019). People, power, change: three pillars of a food sovereignty research praxis, *The Journal of Peasant Studies, 46*:7, 1389–1412, 10.1080/03066150.2018.1512488

De Schutter, O. (2010). *The Right to Food: Interim Report of the Special Rapporteur.* New York, NY: The United Nations.

DeLonge, M. S., Miles, A., & Carlisle, L. (2016). Investing in the Transition to Sustainable Agriculture. *Environmental Science and Policy, 55*, 266–273.

Fakih, M., Rahardjo, T., & Pimbert, M. P. (2003). *Community Integrated Pest Management in Indonesia: Institutionalising Participation and People Centred Approaches.* London: IIED.

Fernando Gomez, L., Rios-Osorio, L., & Luisa Eschenhagen, M. (2013). Agroecology Publications and Coloniality of Knowledge. *Agronomy for Sustainable Development, 33*(2), 355–362.

Giraldo, O. F., & Rosset, P. M. (2018). Agroecology as a Territory in Dispute: Between Institutionality and Social Movements. *The Journal of Peasant Studies, 45*(3), 545–564.

Holt-Giménez, E. (2006). *Campesino a Campesino: Voices from Latin America's Farmer to Farmer Movement for Sustainable Agriculture.* Oakland, CA: Food First Books.

IAASTD. (2008). *Agriculture at a Crossroads: Report of the International Assessment of Agricultural Knowledge, Science, and Technology.* Library of Congress.

IPES-Food. (2016). *From Uniformity to Diversity: A Paradigm Shift from Industrial Agriculture to Diversified Agroecological Systems.* International Panel of Experts on Sustainable Food Systems (IPES).

La Via Campesina. (2017). Peasant Agroecology Schools and the Peasant-to-Peasant Method of Horizontal Learning. https://foodfirst.org/wp-content/uploads/2017/06/TOOLKIT_agroecology_Via-Campesina-1.pdf

Lamine, C. (2018). Transdisciplinarity in Research about Agrifood Systems Transitions: A Pragmatist Approach to Processes of Attachment. *Sustainability, 10*(4).

Levidow, L., Pimbert, M., & Vanloqueren, G. (2014). Agroecological Research: Conforming—or Transforming the Dominant Agro-Food Regime? *Agroecology and Sustainable Food Systems, 38*(10), 1127–1155.

Loevinsohn, M., Sumberg, J., Diagne, A., & Whitfield, S. (2013). *Under What Circumstances and Conditions Does Adoption of Technology Result in Increased Agricultural Productivity? A Systematic Review Prepared for the Department for International Development.* Brighton, UK: IDS.

McCune, N., & Sánchez, M. (2018). Teaching the Territory: Agroecological Pedagogy and Popular Movements. *Agriculture and Human Values, 36*(3), 595–610.

Méndez, V. E., Bacon, C. M., Cohen, R., & Gliessman, S. R. (2015). *Agroecology: A Transdisciplinary, Participatory and Action-Oriented Approach.* Roca Baton: CRC Press.

Mier y Terán Giménez Cacho, M., Giraldo, O. F., Aldasoro, M., Morales, H., Ferguson, B. G., Rosset, P., et al. (2018). Bringing Agroecology to Scale: Key Drivers and Emblematic Cases. *Agroecology and Sustainable Food Systems, 42*(6), 637–665.

Pimbert, M. P. (2018). Democratizing Knowledge and Ways of Knowing for Food Sovereignty, Agroecology and Biocultural Diversity. In M. P. Pimbert (Ed.), *Food Sovereignty, Agroecology and Biocultural Diversity. Constructing and Contesting Knowledge* (pp. 259–321). London: Routledge.

Rosset, P. M., Sosa, B. M., Jaime, A. M., and Lozano, D. R. (2011). The Campesino-to-Campesino agroecology movement of ANAP in Cuba: social process methodology in the construction of sustainable peasant agriculture and food sovereignty. *J Peasant Stud* *38*(1), 161-191. https://doi.org/10.108 0/03066150.2010.538584.

Santos, B. d. S. (2015). *Epistemologies of the South: Justice Against Epistemicide*. New York: Routledge.

Val, V., Rosset, P. M., Zamora Lomelí, C., Giraldo, O. F., & Rocheleau, D. (2019). Agroecology and La Via Campesina I. The Symbolic and Material Construction of Agroecology Through the Dispositive of "peasant-to-peasant" Processes. *Agroecology and Sustainable Food Systems, 43*(7–8), 872–894.

Vanloqueren, G., & Baret, P. V. (2009). How Agricultural Research Systems Shape a Technological Regime that Develops Genetic Engineering But Locks Out Agroecological Innovations. *Research Policy, 38*(6), 971–983.

Visvanathan, S. (2005). Knowledge, Justice and Democracy. In M. Leach, I. Scoones, & B. Wynne (Eds.), *Science and Citizens: Globalization and the Challenge of Engagement*. London; New York Zed Books.

Vogl, C. R., Kilcher, L., & Schmidt, H. (2005). Are Standards and Regulations of Organic Farming Moving Away from Small Farmers' Knowledge? *Journal of Sustainable Agriculture, 26*(1), 5–26.

Waldmueller, J. M. (2015). Agriculture, Knowledge and the 'colonial matrix of power': Approaching Sustainabilities from the Global South. *Journal of Global Ethics, 11*(3), 294–302.

Woodley, E., Crowley, E., de Pryck, J. D., & Carmen, A. (2006). Cultural Indicators of Indigenous Peoples' Food and Agro-Ecological Systems. SARD Initiative commissioned by FAO and the International India Treaty Council.

CHAPTER 6

Domain C: Systems of Economic Exchange

Abstract In this chapter we examine the importance of systems of economic exchange for agroecology. These include the practices and processes by which agricultural products move from producers to various
users and by which agri-food producers acquire inputs that cannot be produced on the farm. We review the importance of traditional systems of
exchange (such as informal markets and barter systems), subsistence (or
family and community self-provisioning) and 'nested markets' that are
embedded in democratic social relations for agroecology. These markets
thicken networks of solidarity and relations of reciprocity in territories.
Nested markets value the ecological, social, economic and political functions and outputs of agroecology and support the development of trust-
based networks. Regrettably, mainstream food markets favour large
volumes and standardization and exclude most agroecological producers.

Keywords Nested markets • Traditional markets • Corporate power •
Global food system • Subsistence

We use the term *systems of economic exchange* (or in shorthand: *systems of
exchange*) in food and farming to mean the practices and processes by
which agricultural products move from producers to various users and by
which agri-food producers acquire inputs that cannot be produced on the
farm. Systems of exchange are thus "the rules-based exchanges of value in

specific contexts where the rules can come from public regulations, private contracts, civic norms or cultural customs" (FAO 2016). They include both formal market mechanisms and informal exchange between agricultural producers of seeds, livestock breeds, labour and more. The extent to which these systems of exchange are accessible, fair, profitable and fulfilling for food producers helps to determine the quality of agroecological transformations.

Agroecology is not anti-trade or against markets *per se*. To be viable, however, it requires systems of exchange that differ starkly from the capitalist, corporate-led systems of exchange that pervade the dominant regime. The existence of appropriate and robust systems of exchange, including different types of markets, state provisioning, barter, gifts and self-sufficiency, are all important enablers of agroecology. Longstanding traditional systems of exchange and the creative construction of newer 'alternative food systems', relations and markets represent a key opportunity for agroecological transformations.

ENABLING CONDITIONS

Agroecological production is based on the integration of a diversity of crops and of livestock; it thus relies on forms of economic exchange compatible with small volumes of many different farm products and local diets. By sustaining a diversity of domesticated and wild foods, agroecological practices themselves are an important enabler of systems of exchange at scales from farm plots to the wider landscape and the commons. Farmers' agroecological practices enhance available dietary diversity by creating micro-environments for growing many different crops and livestock on farms and neighbouring landscapes as well as on the commons—grasslands, forests, wetlands. In addition, these practices sustain key ecological functions at different spatial scales, such as pollination, natural pest control, waste decomposition, water filtration and carbon sequestration (IPBES 2019). These so-called environmental goods and services sustain the material basis of systems of economic exchange important for food and livelihood security.

To support systems of exchange that advance agroecological transformations, it is important to value and build on existing community networks and cultures. Traditional systems of exchange (such as informal markets and barter systems) that have evolved within traditional communities, ecosystems and culture are, although undervalued, a good basis for enabling systems of exchange for agroecology. For example, wild resources found on farms and common lands are often incorporated into agroecological systems. Wild edible plants and animals are particularly important to indigenous people's food and livelihood security as well as that of the rural poor, women and children, especially in times of stress such as drought, shifts in land and water availability or ecological change. With much less access to land, capital and labour, these groups rely on systems of exchange involving wild diversity.

Also key for agroecology are new markets, networks and economic processes that are embedded, or 'nested', in local territories and social relations, for example around definitions of food quality that are mutually agreed by producers and consumers (Jan Douwe van der Ploeg et al. 2012). The Beijing County Fair in China is one example of such new nested markets (Box 6.1). Most commonly, nested markets remove intermediaries as much as possible and are oriented towards direct connections between producers and consumers that build mutual understanding and new solidarities. The Food and Agriculture Organization of the United Nations (FAO) (2018) found that in nested markets, actors are "recapturing value through direct contact, but also through a diversification of their market channels". Nested markets recognize and promote the multiple benefits of agroecological food production—biodiversity, human and ecological health and natural resource management, for instance—which are otherwise undervalued. They also accommodate the diversity of outputs generally produced in agroecological systems, allow for local self-determination and meet the material needs of food producers. This often makes nested markets more attractive for agroecological food producers than conventional markets and global value chains.

Nested markets exist in many forms and under many names. For example, 'alternative food networks' broadly include newly emerging networks of, and relations between, producers, consumers and other actors that embody alternatives to the more standardized industrial systems of food exchange (Kneafsey and Holloway 2008). Some examples of nested market arrangements include participatory guarantee systems, restaurants purchasing food directly from farms, vegetable boxes, farm shops, self-harvest fields and public food procurement (e.g. in university, government and hospital cafeterias). Community-supported agriculture (CSA) is another such arrangement currently on the rise. The international CSA network Urgenci, with members on every continent, defines CSA as "local solidarity-based partnerships between producers and consumers" centred on trust and shared risk.

Building nested markets for agroecology is a case of step-by-step processes based on local resources, in which additional assistance from the state may play a strategic role (Jan Douwe van der Ploeg et al. 2012). Crucial steps in constructing agroecological markets include the diversification of relations and channels (such as through new partnerships with restaurants, educational establishments and consumer groups), resolving post-harvest conservation and storage problems, developing innovative small-scale processing of traditional varieties and carrying out active promotion of these initiatives. The latter often happens by strategically positioning products and creating awareness among consumers, mainly through media, personal communication, farm visits, local events and education. Nested markets thus can have a positive impact on social cohesion, the economic vitality of territories and carbon footprints. They "counter distance with proximity, artifice with freshness, anonymity with identity and genuineness, standardization with diversity and inequality with fairness" (Jan Douwe van der Ploeg et al. 2012).

However, nested markets in some cases replicate the extractive, competitive and exclusionary dynamics and relations of the dominant food system. Based on a heterodox view of economics, the framework for these markets argues that they "coexist with other (conventional) markets and struggle with these for space and legitimacy", and "constitute concrete spaces of interaction between specific actors, which are constructed and reproduced within the conventional markets, that is, within the capitalist mode of production" (Sonnino and Marsden 2006). The politics in some farmers' markets and CSAs, for instance, have been found to be driven as much by profit-seeking and individualism as by logics of solidarity and trust (Hinrichs 2000) or to echo the exclusionary dynamics underlying racial capitalism (Slocum 2007).

Nested markets are vital, but it is important to view them critically and to question their political underpinnings so that they can more effectively foster agroecological transformations. Some forms of such markets are more explicitly opposed to capitalist and extractive economies, for example solidarity economics, de-growth, and eco-feminist, indigenous and anarchist economics.

In agroecology, not only products but also cultural traditions, ideas, visions and knowledge are exchanged. As Stephen Sherwood et al. (2018, p. 5) note, an agroecological market is "a site of social creativity where people situate and territorialize their abilities to affect and be affected", allowing them to shape their own socio-material conditions. The authors illustrate this through a case study of the Carcelen Agroecology and Solidarity Fair in Quito. While state institutions tried to enforce official norms and standards around production, hygiene and price, participants in the fair were "renewing a sense of self and collectivity". Through this they generated relationships focused not only on a need for calories and food security but also on new values connected to cultural expression, health, environmental sustainability and a sense of community. This is one of many ways in which an agroecological approach may reveal the first stirrings of new "regimes"—"food from somewhere" as opposed to corporate "food from nowhere" (McMichael 2009).

Labelling has been promoted as another mechanism for upscaling and securing markets for sustainable food. While third-party labels and certificates have indeed provided important support for the scaling up of different approaches to sustainability in agriculture, such as organic agriculture and fair trade, the mechanism is contested. For producers who want to participate in certification schemes, problems often arise in relation to cost or demands to conform to externally agreed standards that may have little to do with agroecology. If people are urged to trust a label rather than engage, discern and participate in building local food systems, it can reduce citizens to passive consumers and effectively decouple place from production. So, while labelling may have some role to play in enabling systems of exchange for agroecology, a critical question remains: *who* is responsible for developing, implementing and controlling standards and evaluating which are necessary?

Alternatives to third-party labels exist. To ensure a certain level of food safety and quality while not losing control over their production system, producers in countries like China, France, India and Italy have come together to collectively agree on production methods and standards. These autonomous mechanisms are called participatory guarantee systems (PGSs): locally focused quality assurance systems in which producers self-certify, in some cases in collaboration with consumers (for an example, see Box 6.1).

PGSs are the most widely recognized alternative food certification systems. They are built on a foundation of trust, social networks, knowledge exchange and local control. They keep the costs of certification low. They also respond to the need for clarity on what 'agroecological' means, bringing agroecological actors together in territories to negotiate its meaning as it applies to particular contexts. PGSs can also challenge the assumptions of the dominant regime that underlie third-party certification, such as the prioritization of export-oriented production and the idea that only formally trained experts can make valid assessments of quality. As local institutions for collective decision-making, PGSs can therefore be considered a tool for strengthening innovation in agroecology, for challenging the dominant regime in food and farming and for moving to a commons model and away from commodification of agriculture and its products (Vivero-Pol et al. 2018).

Box 6.1   The Beijing County Fair—Building a Commitment to Sustainable Food

Fig. 6.1   CSA members of Little Donkey farm (Beijing, China) harvesting carrots (*Photo credit*: Jan Douwe van der Ploeg)

The Beijing County Fair in China is an example of a nested market that supports agroecology transitions. It was first organized by local consumers and artists in 2010, and by 2015 it was run by 11 full-time staff. Within a decade, it has developed into the most active and influential ecological farmers' market in China. One of its

(*continued*)

Box 6.1   (continued)

managers notes that in 2017 her team organized 154 markets, each time involving about 20 small- to medium-scale ecological farms and 10 smallholder food processors. Other than farmers' markets, the same team also runs 2 grocery stores and an online shop, selling the produce of over 70 farms and 20 food processors' facilities.

The Fair has rebuilt trust between individual food producers and consumers and has developed the trust of consumers in institutions (Wang et al. 2015). One of the Fair's key tools in this context is a PGS. In 2014, the Fair started to experiment with a PGS by developing a farm information form completed by about 30 farms and checked during farm visits, as well as regulating and increasing the frequency of farm visits (Jiang 2015). The form is used to holistically evaluate a farm in terms of technical practice and social aspects such as ownership structure, employment and marketing approaches. Transparency is key to participation in the Fair: the forms are displayed at the farmers' market and also available online.

The Fair is the PGS pioneer in China, but not the only body to adopt it. At the start of 2018, 18 farmers markets (including the Fair), social enterprises and buyer groups across China established a PGS network called 'Clover' to enable collective learning on standards, joint farm visits and communication activities.

Source: Xu Ye and Mindi Schneider, the International Institute of Social Studies (ISS), The Hague, Netherlands

When food safety regulations and validation processes are tailored to small-scale agricultural producers using an agroecological approach, agroecological systems of exchange inevitably benefit. Conversely, rigid and uniform rules on, for example, food safety and plant disease control can severely limit the circulation of artisanal products of small-scale producers and often fail to improve food safety (McMahon 2013). To ensure consistent quality in organic food, legislation and governmental standards have been established for production, processing, trading, monitoring and certification—for example, the European Council Regulation on Organic Farming No. 202, Brazil's organic farming legislation of 2003 and Japan's Agricultural Standards for Organic Agricultural Products and Their Processed Foods.

Some of these regulations, however, were constructed for large-scale farming and processing and could undermine the specific production model of small-scale agroecological producers. For example, organic food production rules and certification rarely take into account the proximity of production and consumption as a safety feature for the nutritional quality of foods (Vogl et al. 2005). For agroecology to thrive, regulatory mechanisms for food safety and quality must allow for regional definitions while supporting small-scale producers' knowledge and socio-technical experiments in sustainability and resilience.

Many actors around the world have called for regulatory, financial and infrastructure state support for markets for agroecology. Indeed, governments can play enabling roles. In light of the broadly recognized human right to food, food cannot be conceived as a commodity like any other. It is thus essential that states intervene in markets. For example, state support was essential for the development of four different types of markets for agroecology in China: the export-oriented market for organic produce; the domestic market for certified food; the localized market for traditional agriculture and typical regional products; and markets for agro-tourism. While many of these markets started off as experiments by farmers, after learning and adjustment they were integrated into government programmes. Each now plays a distinct role in supporting agroecology (Ye et al. 2010).

There are various forms of government support for systems of exchange. One is through public food-procurement programmes such as the Program for Food Acquisition from Family Farming (PAA) in Brazil (Box 6.2). Or states can lend financial, logistical or promotional support to markets for agroecology and thus increase their visibility and viability. A key role for governments here is establishing infrastructure that overcomes impediments in transportation and information networks, for example by building cold-storage systems for fresh fruits and vegetables.

So, the role of governments in promoting markets for agroecology can be key. However, when markets are constructed in a non-participatory manner, they may become counterproductive, as barriers to inclusion, bureaucracy, paperwork and costs may emerge. Moreover, care must be taken that these markets continue to support diversified agroecological food production, especially when the market seems to be shifting to larger volumes or towards export. Similarly, in terms of nutrition and food security, policies to enhance agroecology for sustainable food systems must promote production for household consumption over that for commercial interests.

Box 6.2   Public Food Procurement as a Motor for Agroecology in Brazil

Brazil's Program for Food Acquisition from Family Farming (PAA) was established in 2003 as part of former president Luiz Inacio Lula da Silva's Zero Hunger Strategy. It has a dual objective: to bring quality food to the socially most vulnerable sectors of society *and* to strengthen family farmers, even the most impoverished. Notably, the PAA has stimulated crop diversification and helped to open new marketing channels. With the same budget, it has also had positive impacts in other sectors, such as biodiversity conservation, public health and addressing climate change. For Brazilian social movements, the PAA has been the most innovative and effective public policy for agroecology.

Moreover, since the 1940s Brazil has been running the National School Feeding Program (PNAE), explicitly aimed at creating an institutional market for Brazilian agricultural producers. Since 2009, the PNAE requires that 30% of purchases come from local family farmers, offering a price premium for agroecologically produced food.

The PAA followed an upward path for over a decade. By 2016, it had reached sales of R$850 million (approximately 150 million euros) buying and distributing more than 297,000 tons of food from 380 different products in all the Brazilian states, and benefiting approximately 185,000 farmers' families. This was possible because the PAA involved more than 24,000 social organizations that worked to help families in situations of social vulnerability. In that same year, however, brutal budget cuts began.

Now, in 2020, the PAA budget is reduced to less than R$100 million. The procedures have become very bureaucratic, making participation of the poorest farming families extremely difficult (Oldekop et al. 2015). Social movements are currently, in the midst of the COVID-19 pandemic, organizing for PAA to be revitalized and the PNAE to be improved. Their goal is to resume the original modalities of the PAA programme and increase the budget to R$1 billion by the end of 2020.

Source: Prepared in collaboration with Paulo Petersen, AS-PTA, Brazil

In addition to the 'downstream' side of systems of exchange (i.e. moving goods from producers to users and consumers), agroecology also demands appropriate upstream systems of exchange. The majority of external, capital-intensive inputs need to be gradually displaced by knowledge-intensive practices based on natural processes such as on-farm production of organic fertilizers, the use of natural processes for pest control, intercropping and soil management. These have reduced farmers' dependence on a host of industrial-chemical inputs and their levels of debt. In one example, savings from lower expenses on farm inputs allowed 386 out of 487 households surveyed in Andhra Pradesh, India, to reclaim their mortgaged farmland (Gregory et al. 2017).

There may still be inputs that farmers cannot derive on the farm but need to acquire from other producers through dynamic exchange of seeds, breeding stock, feed, labour, nutrients and tools. These systems of exchange may consist of formal market-based mechanisms or informal relations. Community seed collecting, practised in regions from Asia to Africa, is one such informal system, involving the exchange and systematic sharing of seeds as well as arrangements to exchange manure and feed. Such initiatives are enabled in contexts where civil society networks are developing open source seed systems (Montenegro de Wit 2017), where there is an active movement to reject biopiracy and genetically modified seeds and where peasant seed networks already exist and are being defended (Peschard and Randeria 2020). These points drive home yet again how important power and politics are in the development of agroecological networks.

Another inspiring example is rooted in the idea that farmers themselves are innovators. In the network of L'Atelier Paysan in France, farmers collaborate with engineers, IT specialists and mechanics to develop and exchange tools and self-built machinery for agroecology-based farming. Through the sharing of farm-based inventions, the initiative makes agroecology transdisciplinary. L'Atelier Paysan also engages in farmer-driven projects to build or renovate agricultural buildings. The network's designs for new farm tools and machinery are all disseminated as open source materials, and it runs courses and produces educational materials to share skills and ideas. In these ways, L'Atelier Paysan builds an upstream system of exchange that affirms the principle of technological sovereignty within and between territories.

DISABLING CONDITIONS

One of the most significant barriers to developing agroecology is the absence, or erosion, of appropriate systems of exchange, coupled with the growth of specialized, export-oriented value chains. These mainstream food markets generally demand large volumes of product and standardization, reinforced by policies that emphasize economies of scale, strategic export commodities and integration into global value chains, which many agroecological producers cannot, or opt not to, engage in (IPES-Food 2016; van der Ploeg 2018).

There are many reasons why they don't. Because agroecological approaches focus on crop and genetic diversity, farmers using the system may only rarely produce sufficient quantities of uniformity in single crops to solely participate in export markets and global value chains. Further, commodity prices are often at or below the cost of production. This provides clear benefits to agribusinesses in processing and retail, for instance, but it traps small-scale farmers in cycles where they must "go big or get out"—specialize or be excluded from export markets (Howard 2016). In addition, the current drive to harmonize food safety standards across the world often favours multinational capital and marginalize local small-scale producers, yet creates systemic "un-safety, poor health and a future of food insecurity for many" (McMahon 2013).

Thus, globalized market arrangements do not work well for agroecology. The prices do not reflect the costs, and important non-market values central to agroecological principles are driven out—equity, shared social welfare, solidarity, kinship, reciprocity, culture and traditions among them. An example, described by Alexander Day and Mindi Schneider (2017), shows how the contemporary political economic context in China, which pushes intensified modernization, has compelled agroecological networks to follow the same market logic as state policy-makers—specifically to "focus on niche marketing to the urban middle class, without seeking to transform rural social relations" (Day and Schneider, 2017 p. 1223). These lock-ins pose challenges to markets for agroecology, such as an inability to respond to rising demand because of inconsistent levels of agroecological production, lack of adequate logistics for distribution, low consumer consciousness, limited public sector support and unfair price competition (FAO 2018).

Against all this stands the fact that a minority of the world's food is directly exchanged in global markets: only 12–17% of the total volume crosses an international border between production and consumption

(Chappell 2018, p. 204n8). Many states and policies, driven by concerns about food security, attempt to change this and explicitly prioritize the integration of small-scale food producers into global markets rather than encourage the development of diverse local markets.

But such efforts to make global value chains more 'inclusive' tend to benefit only a small number of farmers worldwide—10% at most—who tend to be well off, educated, strongly oriented towards commercial agriculture and living close to urban areas and infrastructure (Seville et al. 2011). On the consumer side, international trade has mainly benefited wealthy consumers in high-income countries while marginalizing communities in low-income countries who continue to be unable to afford the diversity available on global markets. In Bangladesh, the commercialization of agriculture and the continued forced integration of farmers in the market economy regime are considered to be at least partly responsible for today's high rates of malnutrition among rural people (Misra 2017).

The global overproduction of food and concomitant decline in prices typically harm farmers' livelihoods. Farmers will usually increase production to make up for lower prices for each unit they produce (Chappell 2018, pp. 42–44). In practice, this means that producers are often reluctant or unable to get off this 'treadmill' and may be deterred from shifting to agroecological practice. But it is immensely profitable for corporations, as they are able to sell ever more inputs and buy ever-cheaper agricultural products (Chappell 2018). This in turn helps to lock-in the current regime and block transition, as farmers are often encouraged to adopt new technologies in order to boost production. Another problem with global overproduction is that it forces producers to raise crops or livestock months before they know what the selling price will be.

Markets that provide inputs for agriculture, aided by schemes subsidizing external inputs, pose hurdles for agroecological transition. The concentration and consolidation of these markets has been called "one of the most pressing concerns" related to agricultural industrialization (Hendrickson et al. 2017). Here, again, large corporations make significant profits while pushing farmers into growing resource-intensive, environmentally destructive monocultures for very low prices, often below production cost. The cost of external inputs is a major burden for producers, who turn to subsidy schemes; they then often accelerate and increase their use of fertilizers, pesticides, commercial seeds, non-locally adapted livestock genetics and imported feed. Paying for inputs reduces profit margins, which may trigger a need for credit and risk insurance. (This also happens with livestock production that is dependent on costly external inputs such as feed, medicine or capital-intensive installations such as stables.) As with overproduction and its impact on farmers, a cycle of debt, consolidation and industrialization can result (Chappell 2018; Howard 2016).

To enable farmers to access external inputs, many countries have established public subsidy programmes. A 2016 study by the African Centre for Biodiversity on the effects of state-led farm input subsidy programmes in ten countries in southern Africa found these to be largely ineffective, as a result of grabbing by elites and diversion, for example through theft or sale by beneficiaries (Africa Centre for Biodiversity 2016). According to the study, the subsidies' direct contributions to higher yields and reduced food prices failed to directly benefit the poor and most vulnerable, who are mostly women. Importantly, the input subsidy programmes increase rural communities' dependency on external inputs, impeding any move to agroecology.

Removing such government subsidies for agro-industrial inputs can eliminate perverse incentives that keep farmers hooked on agro-industrial networks. For example, a programme launched in 2003 by the government of Sikkim state in India reduced subsidies for agrochemicals by 10% each year. By 2007–2008, they were eliminated, and by 2009, the sale of all agrochemical products was phased out (Gregory et al. 2017). In concert, the state aimed to support the development of a bio-input industry and to develop markets for the organic products of Sikkimese agriculture; however, unfortunately many of these policies were ill-conceived and in practice served to undermine agroecology (Meek and Anderson 2020; see Box 10.1 in Chap. 10).

REFERENCES

Africa Centre for Biodiversity. (2016). *Farm Input Subsidy Programmes (FISPs): A Benefit for, or the Betrayal of, SADC's Small-Scale Farmers?*
Chappell, M. J. (2018). *Beginning to End Hunger: Food and the Environment in Belo Horizonte, Brazil, and Beyond*. Oakland: Univ of California Press.
Day, A. F., & Schneider, M. (2017). The End of Alternatives? Capitalist Transformation, Rural Activism and the Politics of Possibility in China. *The Journal of Peasant Studies, 45*(7): 1221–1246.
FAO. (2016). *Innovative Markets for Sustainable Agriculture. How Innovations in Market Institutions Encourage Sustainable Agriculture in Developing Countries* (A. Loconto, A.S. Poisot, & P. Santacoloma, Ed.). Rome: Food and Agriculture Organization of the United Nations/Institut national de la recherche agronomique [French National Institute for Agricultural Research].

FAO. (2018). *Constructing Markets for Agroecology—An Analysis of Diverse Options for Marketing Products from Agroecology.* Rome: FAO.

Gregory, L., Plahe, J., & Cockfield, S. (2017). The Marginalisation and Resurgence of Traditional Knowledge Systems in India: Agro-Ecological 'Islands of Success' or a Wave of Change? *South Asia-Journal of South Asian Studies, 40*(3), 582–599.

Hendrickson, M. K., Howard, P. H., & Constance, D. H. (2017). *Power, Food and Agriculture: Implications for Farmers, Consumers and Communities.* Division of Applied Social Sciences Working Papers, University of Missouri College of Agriculture, Food & Natural Resources.

Hinrichs, C. C. (2000). Embeddedness and Local Food Systems: Notes on Two Types of Direct Agricultural Market. *Journal of Rural Studies, 16*(3), 295–303.

Howard, P. H. (2016). *Concentration and Power in the Food System: Who Controls What We Eat?* London: Bloomsbury Academic Publishing.

IPBES. (2019). *Global Assessment Report on Biodiversity and Ecosystem Services.*

IPES-Food. (2016). *From Uniformity to Diversity: A Paradigm Shift from Industrial Agriculture to Diversified Agroecological Systems.* International Panel of Experts on Sustainable Food Systems (IPES).

Jiang, Y. (2015). The Chinese CSA Movement Gaining Momentum, in Sustainable Agriculture in China: Land Policies, Food and Farming Issues. *China-Program of the Stiftung Asienhaus.* Cologne, Germany.

Kneafsey, M., & Holloway, L. (2008). *Reconnecting Consumers, Producers and Food: Exploring Alternatives.* Berg Publishers.

McMahon, M. (2013). What Food Is to Be Kept Safe and for Whom? Food-Safety Governance in an Unsafe Food System. *Laws, 2*(4), 401–427.

McMichael, P. (2009). A Food Regime Genealogy. *Journal of Peasant Studies, 36*(1), 139–169.

Meek, D., & Anderson, C. R. (2020). Scale and the Politics of the Organic Transition in Sikkim. *India. Agroecology and Sustainable Food Systems, 44*(5): 653–672.

Misra, M. (2017). Moving Away from Technocratic Framing: Agroecology and Food Sovereignty as Possible Alternatives to Alleviate Rural Malnutrition in Bangladesh. *Agriculture and Human Values, 17*(3), 594–611.

Montenegro de Wit, M. (2017). Beating the Bounds: How Does 'open source' Become a Seed Commons? *The Journal of Peasant Studies, 46*(1), 44–79.

Oldekop, J. A., Chappell, M. J., Peixoto, F. E. B., Paglia, A. P., do Prado Rodrigues, M. S., & Evans, K. L. (2015). Linking Brazil's Food Security Policies to Agricultural Change. *Food Security, 7*(4), 779–793.

Peschard, K., & Randeria, S. (2020). 'Keeping Seeds in Our Hands': The Rise of Seed Activism. *The Journal of Peasant Studies, 47*(4), 613–647.

Seville, D., Buxton, A., & Vorley, B. (2011). *Under What Conditions Are Value Chains Effective Tools for Pro-Poor Development?* International Institute for Environment and Development/Sustainable Food Lab.

Sherwood, S., Arce, A., & Paredes, M. (2018). Affective Labor's 'unruly edge': The pagus of Carcelen's Solidarity & Agroecology Fair in Ecuador. *Journal of Rural Studies.* Vol. 61: 302-313. https://doi.org/10.1016/j.jrurstud.2018.02.001

Slocum, R. (2007). Whiteness, Space and Alternative Food Practice. *Geoforum, 38*(3), 520–533.

Sonnino, R., & Marsden, T. (2006). Beyond the Divide: Rethinking Relationships between Alternative and Conventional Food Networks in Europe. *Journal of Economic Geography, 6*(2), 181–199.

van der Ploeg, J. D. (2018). *The New Peasantries: Rural Development in Times of Globalization.* Earthscan Food and Agriculture.

van der Ploeg, J. D., Jingzhong, Y., & Schneider, S. (2012). Rural Development Through the Construction of New, Nested, Markets: Comparative Perspectives from China, Brazil and the European Union. *Journal of Peasant Studies, 39*(1), 133–173.

Vivero-Pol, J. L., Ferrando, T., De Schutter, O., & Mattei, U. (2018). *Routledge Handbook of Food as a Commons.* Routledge.

Vogl, C. R., Kilcher, L., & Schmidt, H. (2005). Are Standards and Regulations of Organic Farming Moving Away from Small Farmers' Knowledge? *Journal of Sustainable Agriculture, 26*(1), 5–26.

Wang, R. Y., Si, Z., Ng, C. N., & Scott, S. (2015). The Transformation of Trust in China's Alternative Food Networks: Disruption, Reconstruction, and Development. *Ecology and Society, 20*(2), 19.

Ye, J., Rao, J., & Wu, H. (2010). Crossing the River by Feeling the Stones: Rural Development in China. *Rivista di Economia Agraria, 65*(2), 261–294.

CHAPTER 7

Domain D: Networks

Abstract In this chapter we examine how local organizations, affinity groups and the formal and informal networks they form provide the basis for the collective, coordinated actions needed for agroecological transformation at different scales. Civil society-driven networks are crucial because they facilitate a kind of cooperation that cannot be generated by the market or the state. On the other hand, the absence of appropriate networks can substantially limit agroecological transition, for example where political dynamics undermine or weaken the development of networks for collective action. Another disabling dimension of this domain is the compartmentalization of networks (e.g. by commodity group), which is a contradiction to the holism of agroecology. Perhaps most challenging is the growing individualization of society that is creating a growing barrier to cooperativism.

Keywords Local organizations • Social movements • Farmer organizations • Collective action

As we have seen, local groups and multi-actor networks rooted in civil society are pivotal for making agroecological transformations possible. The other key domains—knowledge, systems of exchange, discourse, efforts to address inequity—are all generated through social organization and networks acting at different scales. Local organizations, affinity groups and the formal and informal networks they form provide the basis for the

C. R. Anderson et al., *Agroecology Now!*,
https://doi.org/10.1007/978-3-030-61315-0_7

collective, coordinated actions needed for agroecological transformation at different scales (Pimbert 2005, 2009).

While experimentation and innovation on farms are generally considered critical agroecological 'field laboratories', social organization in networks increase the reach, depth and potential of such innovations. In this sense, Mier y Terán Giménez Cacho et al. (2018) argue that *organicidad*, or the degree of organization, is the "culture medium" on which agroecology grows. This may explain why agroecological farmers tend to have closer engagement with networks (e.g. of fellow farmers, academics and NGOs) than conventional farmers (Teixeira et al. 2018). Indeed, there is widespread evidence on the importance of grassroots networks in developing social innovation and alternative approaches to the dominant regime in food, farming and beyond. Such networks are key for connecting 'islands of success', building systems of exchange that enable learning and the co-production of knowledge, share labour and resources, and work collectively in the political arena.

As we will detail here, while mainstream agriculture is advancing through processes of individualization and de-territorialization, collective organization in networks and organizations is supporting the transition to agroecology. Through that process, local alliances and movements are better able to adaptively manage agroecosystems and landscapes; coordinate human skills, knowledge and labour to generate economic wealth and exchanges in multifunctional food systems; and support shifts in governance of food systems as well as facilitate coordinated action for policy and institutional change.

ENABLING CONDITIONS

Innovations in agroecology do not automatically spread to other socio-environmental contexts. They depend on networks of people with the agency to do so. As such, networks enable innovative practices to mature, to reach a greater diversity and numbers of actors such as food consumers, to access resources and to link up to power-holders, which can create opportunities for influencing the regime. Frank Geels and Jasper Deuten (2006) argue that innovation networks generally emerge first with isolated experiments, later moving on to aggregate lessons learned from these through exchanges between actors such as policy-makers and scientists. This phase is crucial for stabilizing socio-technical niches for innovation such as agroecology. After that, intermediaries such as local authorities and relatively stable networks form, bringing together knowledge from local initiatives as well as new actors and activities.

Civil society engagement in these networks is critical because it facilitates a kind of cooperation that cannot be generated by the market or the state (van der Ploeg 2018). 'Endogenous' farmers' networks, for example, have helped increase agricultural diversity and build knowledge, skills and cooperation needed to improve nutrition among farmers (Deaconu et al. 2019). Indeed, self-organization in agroecology is an effective but neglected force for spreading practice, knowledge and participation. In this respect, Schot and Geels (2008) point out that to be effective, networks must be *deep*: members should also be able to generate commitment and resources within their own organizations and networks. Critical pedagogy, horizontal learning, transformative learning and intercultural dialogue are important in this respect because they build trust across boundaries, nurture a collective capacity for critically reflexive practice, address inequity and solve conflicts.

At the same time, and somewhat paradoxically, 'weak' network ties and fluid relationships allowing collaboration, as opposed to close bonds shared among like-minded people, can be important in generating and sustaining change (Nelson et al. 2013). The robustness of networks for agroecology depends on the material and social resources they are able to develop, wield and maintain and the way these are combined to mutually reinforce each other.

Within networks, certain capacities can develop: to organize, to learn and to continuously improve practice by distancing from extractive markets, upstream and downstream while collectively constructing markets for agroecologically produced goods (see Chap. 6 on the systems of economic exchange domain). Other such capacities include the ability to develop a shared sense of place and identity, the commitment to collaborate towards a common goal, and processes of critical education and peer-to-peer knowledge building (Anderson et al. 2019; see also Chap. 5 on the knowledge and culture domain). The more numerous the connections between these resources, the more attractive, accessible and useful the networks become for agroecology transformations.

Experimentation by, within and between networks can also focus on institutional issues such as markets or governance. New formal or informal institutions for local regulation and governance are often created in networks to fill institutional voids, leading to salutary developments such as innovative public policies. When such policies prove workable, they can in turn expand the reach of such territorial governance and promote institutional collaboration between state and non-state actors. For example, Brazil's Ecoforte programme explicitly recognizes and financially supports the role of territorial networks and governance for the promotion of agroecology (González de Molina et al. 2019).

State support can be crucial for the development of new institutions such as common innovation platforms and learning networks. To ensure that local needs and ownership are prioritized, Yoko Kanemasu (2008) argues that such support to networks should be based on broad participation across sectors, bringing together agriculture, health, environment and other actors with a stake in food systems.

More generally, networks supporting agroecology are most likely to thrive in contexts where a vibrant civil society is encouraged and nurtured through policies, regulations, norms and institutions that support human rights and bottom-up processes. In Thailand, for example, sustainable agricultural groups such as the Alternative Agriculture Network, established in the 1980s by farmers and local NGOs, were afforded the political space they needed to develop when the government launched major programmes to support sustainable, organic and self-sufficient farming.

Bringing on Board New Actors

The most promising agroecology initiatives are those where grassroots actors reach across divides and organize to get others on board to create new, multi-actor constituencies with common aims and interests (IPES-Food 2016). Involving outsiders in a process of social organization can increase the scope of resources available to them, such as knowledge, access to other networks, political influence and finance. In turn, these can increase the reach of agroecology.

A diversity of organizations and actors with different functions, powers, resources, membership and responsibilities is usually needed in a network to coordinate the variety of activities needed to amplify agroecology. Interlinked organizations provide the broad institutional landscape—and the means for coordinated action—required to manage the dynamic social and ecological complexities in which agroecology-based food systems are embedded, and for the systems themselves (Pimbert 2009).

For example, as described by Leonardo van den Berg et al. (2016), in the municipality of Araponga in the state of Minas Gerais, Brazil, networks that bring together farmer unions, academic groups and NGOs have been an important factor of success for both the development and territorial spread of agroecology and for acquiring financial support, fostering further experimentation and innovation, and obtaining formal legitimacy. This case emphasizes the importance of a step-by-step process for drawing in both new actors and new activities: first land reform, then agroecology. This strengthened and increased the efficacy of the network with every step, thereby deepening the transformational process.

Dynamic alliances for agroecology are also being built at the international scale, such as the Nyéléni network comprising La Via Campesina, the World March of Women and the Network of Peasant Organizations and Agricultural Producers in West Africa (ROPPA). In these, the movements of a diverse range of food producers, consumer constituencies and other actors engage in discussions, exchanges and joint activities to promote agroecology and food sovereignty (see www.foodsovereignty.org).

At a certain point in a network's development, it can be helpful to engage policy-makers, scientists and other institutional actors from within the dominant regime. State officials might join in, particularly those sympathetic to agroecology or who can contribute to its learning processes by applying or 'translating' the official literature and cognitive frameworks often used by state entities (Ortiz et al. 2017). In this way, insights, resources and knowledge from the regime can strategically modify the agroecological niche, while knowledge, discourse, governance and other elements of agroecology may inform and change the dynamics of the regime. Networks can therefore become important spaces for interaction between the agroecological niche and the regime.

However, the involvement of such actors in networks, such as through the injection of financial resources, can also become problematic by subtly leading to gradual co-optation (see Part III). Incentives and support can enable elements of agroecology that most resemble those of the *status quo* and diminish those that are transformative. A study of agroecology niches in South Africa (Metelerkamp et al. 2020) found that "state-led extension services and formal training institutions are of little help to niche pioneers and instead contribute toward the path-dependency of the current food regime".

So, while necessary, working within such networks is also complex and power-laden, posing a risk of depoliticization. Care must be taken regarding the influence of well-resourced and often well-meaning bureaucrats who—like NGOs, scientists, privileged activists and other power-holders—can wield disproportionate influence over the way agroecology scales out and up. Risks of co-optation emerge when the agendas and priorities of more powerful participants change the nature and values of niche innovations and the internal dynamics of networks. Agroecological farmers and others within the niche may end up marginalized. This happened, for instance, when organic agriculture was absorbed into corporate-led chains over the last few decades, undermining the values and transformative potential of the pioneering organic movement on which the practice was founded.

This is a conundrum in agroecology: the influence of the regime can fragment, diminish and even intentionally suppress the knowledge, markets, equity, discourse, access and rights over nature required for agroecology. Substantial resources are then needed to rebuild these domains.

The issue also points at the crucial role of facilitators and coordinators in strengthening local organizations and weaving together effective networks. New actors may find it difficult to embrace the same range of values, expectations, rules, norms and politics that have provided consistency in emergent agroecological initiatives and that underpin their transformative potential. The inclusion of such actors should be facilitated to avoid creating any dependencies on scientists and political parties, for instance, that undermine autonomous, long-term, genuine agroecological transformations (Ortiz et al. 2017).

As decisions are made over whom to admit to networks, the question of equity related to class, gender, caste, religious and race divisions comes to the fore. Agroecological organizations and networks are not always inclusive towards women and the marginalized nor—as we have noted— are they free from manipulation by more powerful actors. They can be plagued by internal inequities and social injustices, with decisions taken by men, landowners, people in 'upper' castes or privileged classes at the expense of the relatively powerless—women, landless farm workers, pastoralists, forest peoples and urban slum dwellers among them. Attempts to build intersectoral and intersectional movements—for example between women's groups, food movements, climate activists and migrant networks—are a promising pathway to solving shortcomings related to equity, gender, social inclusion, race, privilege and entitlement. As Eric Holt-Gimenez and Yi Wang (2011) point out, the political direction of the food justice movement's organizational alliances will be towards either reform or transformation, depending on how issues of race and class are resolved. These issues are further elaborated in Chap. 8 on the equity domain.

Various forms of effective formal and informal multi-actor networks for agroecology include community-supported agriculture initiatives and collaborations between groups of agricultural producers and researchers. In Colombia, at territorial and national levels, networks including farmers' associations and families, supportive NGOs, donors, researchers and government entities successfully aggregated, shared and applied lessons and proved key in advancing agroecology and family farming (Ortiz et al. 2017). Food policy councils are an increasingly relevant approach at the municipal and sometimes territorial level, bringing together people from sectors such as food, public health, agribusiness, retail, policy and civil society to develop long-term food-related strategies. Multi-stakeholder platforms, although not without problems, are a critical figure for advancing land tenure governance around the world (Box 7.1).

Box 7.1 Networking Lessons from a Multi-actor Platform for Land Tenure in South Africa

In South Africa, an inclusive and participatory networking approach spurred the government to implement the Voluntary Guidelines on the Responsible Governance of Tenure of Land, Fisheries and Forests (VGGT) to address three national priorities: food security and nutrition, sustainable and equitable natural resource management, and sustainable land reform. At the heart of this collaborative process, which began in 2013, is a multi-stakeholder platform where dialogue and consensus-building on priorities could take place. National VGGT workshops and learning events and programmes built the awareness and capacity of different actors in civil society, grassroots organizations and beyond to use the VGGT effectively.

Several lessons can be drawn from this initiative. First, a process involving multiple actors strengthens the consensus around the needed policy and legal reforms: it can generate or increase the political will needed to adopt and implement such reforms. Second, this experience showed the importance of time: to progressively build trust among partners, even within a specific actor group; to reach a common understanding of the situation; and to develop a consensus on the common strategy. Third, it stressed the critical role of a credible, neutral convener and facilitator in the process, not least to address and overcome power asymmetries. Lastly, a core group of people around this facilitator is also needed to keep the momentum and move forward.

Source: HLPE (2018)

DISABLING CONDITIONS

Today's food system is reinforced by a mesh of interrelated market and policy incentives for subsidies, retail, research, international trade and more, as well as social and cultural dynamics that lock farmers into agriculture that is specialized, large scale and highly dependent on external inputs (IPES-Food 2016). This approach can significantly constrain the development of effective networks for agroecological transformation.

While agroecology can be practised by one farmer, a transformative agroecology is deeply collaborative. Yet decades of neo-liberalization have created a worldview in which the individual is the defining unit of economic and political action. Individual choice and freedom in a free market has become culturally and institutionally entrenched. Farmers are seen as entrepreneurs and citizens as consumers, while the role of civil society is depoliticized and degraded.

Mainstream discourse on conservation and development envisages that the number of farmers, fishers and other people engaged in land- or water-based livelihoods will shrink. Since the 1980s, 'get big or get out' has been a common credo. In this view, increased farm size is seen as a prerequisite for individual farmers eager to develop and become more included in the mainstream economy. Encouraging people who cannot (or will not) scale up to move out of agriculture and get jobs in the largely urban-based manufacturing and service sectors is seen as both desirable and necessary—regardless of the social and ecological costs. This dominant discourse on modernity and progress, and its focus on the individual farm, directly encourages and legitimates the active neglect and undermining of networks and social organization for agroecological transformation (this is further discussed in Chap. 9 on the discourse domain).

Another key barrier to developing effective multidisciplinary and multi-actor networks for agroecology is the compartmentalization of different aspects of the food system. Institutions and organizations that focus on one specific aspect of the food system (such as technology, seeds, markets, natural resources, health and consumption) may lose sight of the holistic approach at the heart of agroecology. For example, within farmer organizations, a strong sectoral focus embedded in highly specialized production can also prevent networking between producers—a requirement for the spread of agroecology. Similarly, synergies are often lacking between governmental policy departments, for example those for agriculture, fisheries, forestry, education, health, water or the environment. This can create difficulties for an integrated approach to food systems. This 'pillarization' is also widespread in academic disciplines involved in research on agriculture, rural society or development. That the effect of this tendency is detrimental has long been pointed out by scholars, and has been discussed in detail in Chap. 5 (on the knowledge and culture domain).

Another major obstacle to the development of networks for agroecology has been the broad reluctance by state entities to empower such processes with the resources—budgetary, logistical, legal and political—to put their deliberations into practice. Official recognition has been critical for successful networks, especially in the early stages of transformation. When such resources are not allocated, involvement in networks can lead to exhaustion and disengagement. Perhaps most importantly, there is a need for much more recognition and support for food producers' organizations that enhance the agency, knowledge and identity of small-scale agricultural producers and their rights. This would steer policy and discourse away from their current emphasis on the economic vitality of farming, which dominates in mainstream farming organizations, research centres and policy institutions. The recent recognition of peasants' rights by the UN Human Rights Council may represent an important step forward in the global recognition of farmers' agency.

The development of agroecology networks can also be disabled by a country's political and societal context. In some countries, support for civil society may not exist; in China, the establishment of networks and local organizations with alternative views is discouraged and suppressed (Castella and Kibler 2015). In other countries, the political context can be outright hostile to dissenting social organization. Ellinor Isgren and Barry Ness (2017) describe how civil society organizations in Uganda historically have limited capacity or space for political organizing, fearing consequences such as deregistration, harassment or arrest. This is not unjustified: Uganda's recently passed NGO Act allows for tighter control of civil society. And in September 2017, Ugandan authorities closed down the bank accounts of the international NGO ActionAid, a strong advocate of agroecology and democratization in the country.

References

Anderson, C. R., Maughan, C., & Pimbert, M. P. (2019). Transformative Agroecology Learning in Europe: Building Consciousness, Skills and Collective Capacity for Food Sovereignty. *Agriculture and Human Values, 36*(3), 531–547.

Castella, J.-C., & Kibler, J.-F. (2015). *Actors and Networks of Agroecology in the Greater Mekong Subregion*. AFD.

Deaconu, A., Mercille, G., & Batal, M. (2019). The Agroecological Farmer's Pathways from Agriculture to Nutrition: A Practice-Based Case from Ecuador's Highlands. *Ecology of Food and Nutrition, 58*(2), 142–165.

Geels, F., & Deuten, J. J. (2006). Local and Global Dynamics in Technological Development: A Socio-Cognitive Perspective on Knowledge Flows and Lessons from Reinforced Concrete. *Science and Public Policy, 33*(4), 265–275.

González de Molina, M., Petersen, P. F., Peña, F. G., & Capor, F. R. (2019). *Political Agroecology: Advancing the Transition to Sustainable Food Systems*. Boca Raton: CRC Press.

HLPE. (2018). *Multi-stakeholder Partnerships to Finance and Improve Food Security and Nutrition in the Framework of the 2030 Agenda*. Rome: HLPE.

Holt-Giménez, E., & Wang, Y. (2011). Reform or Transformation? The Pivotal Role of Food Justice in the U.S. Food Movement. *Race/Ethnicity: Multidisciplinary Global Contexts, 5*(1), 83–102.

IPES-Food. (2016). *From Uniformity to Diversity: A Paradigm Shift from Industrial Agriculture to Diversified Agroecological Systems*. International Panel of Experts on Sustainable Food Systems (IPES).

Isgren, E., & Ness, B. (2017). Agroecology to Promote Just Sustainability Transitions: Analysis of a Civil Society Network in the Rwenzori Region, Western Uganda. *Sustainability, 9*(8), 1357.

Kanemasu, Y. (2008). The Impact of Policy Arrangements. In J. D. v. d. Ploeg, & T. Marsden (Eds.), *Unfolding Webs—The Dynamics of Regional Rural Development* (pp. 211–225). Koninklijke Van Gorcum.

Metelerkamp, L., Biggs, R., & Drimie, S. (2020). Learning for Transitions: A Niche Perspective. *Ecology and Society, 25*(1), 14.

Mier y Terán Giménez Cacho, M., Giraldo, O. F., Aldasoro, M., Morales, H., Ferguson, B. G., Rosset, P., et al. (2018). Bringing Agroecology to Scale: Key Drivers and Emblematic Cases. *Agroecology and Sustainable Food Systems, 42*(6), 637–665.

Nelson, E., Knezevic, I., & Landman, K. (2013). The Uneven Geographies of Community Food Initiatives in Southwestern Ontario. *Local Environment, 18*(5), 567–577.

Ortiz, W., Vilsmaier, U., & Acevedo Osorio, Á. (2017). The Diffusion of Sustainable Family Farming Practices in Colombia: An Emerging Sociotechnical Niche? *Sustainability Science, 13*(3), 829–847.

Pimbert, M.P. (2005). Supporting locally determined food systems: the role of local organizations in farming, environment and people's access to food. In: Bigg, T. and D. Satterthwaite (Eds) How to Make Poverty History. London: IIED.

Pimbert, M. P. (2009). *Towards Food Sovereignty. Reclaiming Autonomous Food Systems. Reclaiming Diversity and Citizenship Series.* Coventry: Coventry University.

Schot, J., & Geels, F. W. (2008). Strategic Niche Management and Sustainable Innovation Journeys: Theory, Findings, Research Agenda, and Policy. *Technology Analysis & Strategic Management, 20*(5), 537–554.

Teixeira, H., van den Berg, L., Cardoso, I., Vermue, A., Bianchi, F., Peña-Claros, M., et al. (2018). Understanding Farm Diversity to Promote Agroecological Transitions. *Sustainability, 10*(12), 4337.

van den Berg, L., Hebinck, P., & Roep, D. (2016). 'We Go Back to the Land': Processes of Re-peasantisation in Araponga, Brazil. *The Journal of Peasant Studies, 45*(3), 653–675.

van der Ploeg, J. D. (2018). *The New Peasantries: Rural Development in Times of Globalization*: Earthscan Food and Agriculture.

CHAPTER 8

Domain E: Equity

Abstract In this chapter, we examine how marginalization and inequity—from international policy arenas to the household level and along the intersecting dimensions of gender, age, class and caste, religion, health and race—pose a major barrier to the development of sustainable food systems. The more transformative edges of the agroecology movement are advancing feminist, decolonial and anti-racist approaches that move the analysis from the centres of power to the margins where the hitherto excluded and oppressed are claiming power. Inequity manifests in overt discrimination as well as unequal access to resources and decision-making power at the household or farm level or to markets, credit, knowledge, governance, relations and other resources at the community or territorial level. In the absence of a focus on equity, efforts to advance agroecology risk exacerbating inequity.

Keywords Gender • Feminism • Discrimination • Decoloniality • Oppression

As described in the introduction to this book, today's dominant food system is rooted in corporate power and built on centuries of racist, patriarchal, colonial relations. The result is an unevenly developed global food system that can be seen in different national and local forms.

© The Author(s) 2021
C. R. Anderson et al., *Agroecology Now!*,
https://doi.org/10.1007/978-3-030-61315-0_8

Power in this system is concentrated in the hands of a privileged minority, while the social and environmental "externalities" disproportionately burden already oppressed groups (Holt-Gimenez and Harper 2016). Transformations towards more sustainable and just food systems are thus hampered by dynamics of marginalization and inequity. From international policy arenas to households, the intersecting dimensions of gender, age, class and caste, religion, health and race pose a major barrier to an agroecological transformation of the food system. As Olivier De Schutter and Christine Campeau (2018) have stated, "We cannot sustainably improve how we produce and consume food without addressing questions of power and of inequality." This is, they note, not just about productivity; it also points to a need to "reframe the problem of hunger and malnutrition as a problem of social justice".

Supporting agroecology without directly and actively addressing these issues is likely to either replicate or reinforce inequity. This is why transformative agroecology, with its political and social aspirations and roots in social organization (see Chap. 7 on the networks domain), is one of the most promising pathways for pursuing equity within agriculture and food systems. Agroecology must be an arena for work towards food justice, for decolonializing food (Grey and Patel 2014) and for pursuing feminist approaches to food sovereignty (Soler et al. 2019). Equity is a critical domain of transformation, where agroecology can fulfil its potential as a part of wider calls and movements for social and environmental justice.

A particular focus for those promoting agroecology and food sovereignty is gender inequity, because it intersects with so many other forms of inequity (Mora and De Muro 2018), varying in contexts from caste, social class, sexual orientation and religion to race and age. Inequity manifests in different ways. It can be seen in unequal access to resources and decision-making power (at the household or farm level) or to markets, credit, knowledge, governance, networks and other resources (at the community or territorial level). At the national and global levels, inequity is inherent in the power of the agribusiness sector (Pimbert and Lemke 2018).

Agroecology can be well suited to strengthening equity through a deeply political process, but inequity cannot be undone through agroecological practices alone. Movements centred on issues such as climate change or food security often fail to engage meaningfully with social justice and equity issues (Wretched of the Earth 2019). Without explicit work to address inequities, the mainstreaming, or 'massification', of agro-ecology risks that a minority of actors in the dominant regime become the voice of agroecology while the majority at the margins become further disempowered. In that way, those in positions of power, including researchers like us (see Gómez et al. 2013), are generally part of the lock-in of the dominant regime; however, where 'marginalized' people gain power, transformation is more likely.

Enabling Conditions

Angela Davis, distinguished professor emerita at UC Santa Cruz, once noted: "In a racist society, it is not enough to be non-racist, we must be anti-racist." Much cited in 2020, this commitment can be applied to all forms of structural inequity. An agroecology that contributes to social justice can exist only in communities, territories and societies actively working to dismantle systems of oppression, privilege and inequity. This means prefiguring and modelling more equitable farms, community spaces, relationships, organizations, attitudes and actions. It also means working through collective action—often as a part of social movements—to identify, confront and dismantle inequitable cultures, policies and institutions.

Improved gender equality and the empowerment of rural women can drive various aspects of agroecology, including improved nutrition and increased crop and genetic diversity among others (De Schutter and Campeau 2018). Women often play crucial, but underappreciated or invisible, roles in agroecology as the guardians of seeds and local breeds, with specialized knowledge and skills for preserving and using them for food, feed, spiritual and medicinal purposes. They also often contribute holistic and nutrition-centred perspectives as well as an eye for not just economic but also health, environmental and social needs. Affirming and protecting women's work and insights can thus advance agroecological transformation. It should not, however, become a way to reify gendered differences in roles and knowledge, which often adds to women's workload. Rather, improving equity can be seen as increasing the fluidity of gender roles, allowing improved cooperation, labour division, decision-making, living conditions and governance by directly tackling the many destructive impacts of patriarchal and misogynist dynamics.

Since norms and perceptions that shape gender relations evolve along with changes in the production system (Lambrecht 2016), agroecology itself can be an instrument for promoting gender equity and women's self-empowerment. This emancipatory potential is tied to agroecology's emphasis on local and diversified knowledge, skills and tasks; input independence; and co-creation. In fact, since many agroecology initiatives are led by women, their participation in decision-making at the household and community level is often both an essential prerequisite for and a result of agroecological innovation (Lopes and Jomalinas 2011).

Box 8.1   Understanding Gender Relations and Equity

Gender relations are the rules, traditions and social relationships in wider culture and organizations which together determine how power is allocated, and used differently, by women and men. Inequalities between the power of women and men are primarily caused by structural and institutional discrimination.

The concept of *gender equity* relates to social justice in these relations according to each person's specific needs and possibilities. *Empowerment of women*, on the other hand, implies building critical awareness and agency to transform the structures that produce gender inequalities. Thus, empowerment can be seen as a process of change on the path to greater equity, both at the individual and collective level (following de Marco Larrauri et al. 2016).

In addition, literature is emerging that examines the relations between *gender diversity* and sustainable food and farming while attention to gender diversity is also gaining ground in farmers' organizations (e.g. La Via Campesina 2016).

Thus, gender equity and agroecology can be mutually beneficial. Because learning and knowledge-sharing are at the heart of agroecology, it can provide spaces for women to work in solidarity and gain livelihoods, income and agency at productive, reproductive and community levels (Khadse 2017). In many documented cases, participating in agroecological networks helped women rise out of sometimes violent situations of isolation and affirm their own identity and knowledge (Galvão Freire 2018). In the words of Jan Douwe van der Ploeg (2018, p. 236): "Agroecology needs peasant women and peasant women need agroecology."

Nevertheless, agroecology *per se* does not explicitly address patriarchy and other forms of gender-based inequality that can undermine a socially just process of transformation (de Marco Larrauri et al. 2016). While agroecology's theoretical underpinnings and principles are rooted in the promotion of equity, its practice does not always reflect this. There may be many women involved in agroecology initiatives and social movements, but many may remain hidden as 'wives of farmers' rather than potential leaders (Khadse 2017). As Olga de Marco Larrauri et al. (2016) argue, much agroecology work has not yet incorporated an explicit gender analysis and has thereby permitted the persistence of hidden "internal contradictions" in the farming family (p. 2). Several cases show how, when agroecology efforts are not accompanied by an intentional equity focus, there is a real risk that it adds to women's unpaid care and work burden. Persistent gender inequity may well become an obstacle for the spread of agroecology in turn.

A virtuous interaction between the agroecological movement and the feminist movement is essential to 'de-normalize' gender inequity, along with other inequitable relations and patterns in communities that grow and process food. A feminist perspective on agroecology is useful, as is a critical pedagogy that analyses the condition of women's subordination and pathways to change (Schwendler and Thompson 2017). Explicit efforts must be made to value women's work, empower women politically and address socially constructed gender roles. In terms of women's self-organization, improved access to resources and education regarding agroecological practices and socio-political equity, two things are needed: deliberate, contextualized action and appropriate interventions (FAO 2018; Lopes and Jomalinas 2011). This demands methodological designs and indicators that, first of all, make gender inequity visible (de Marco Larrauri et al. 2016) through reflection and discussion by women on their realities, for instance, to allow it to be addressed. The work of AS-PTA, an NGO in the state of Paraiba, Brazil, provides important lessons in this respect (Box 8.2).

Studies on initiatives that were successful in transforming gender relations in agri-food systems show the key role of iterative, dialogue-based and women-led experimentation with agroecological practices such as diversification, intercropping, nutrition education and marketing innovations (de Marco Larrauri et al. 2016). These strategies often started from a collective reflection by women on their condition, providing a means for them to understand and challenge it using their own agency and collective

> **Box 8.2   How Women Came Out of Isolation Through Agroecology in Brazil**
>
> For over 15 years, AS-PTA, a Brazilian NGO, had been supporting family farmers in developing agroecological innovations. But despite successes, a patriarchal culture remains dominant both within the families and in farmers' organizations in the state of Paraiba. This made women's knowledge, practices and importance for the farm household invisible. It became clear that the inequity between men and women was a barrier to the full implementation of agroecology across the region.
>
> So AS-PTA started to work with rural women in Paraiba. Step by step, the women built a collective identity: 'women farmer-innovators in agroecology'. They accomplished this through meeting, exchanging and reflecting on their realities and work. Making their knowledge visible and explicit motivated many women to expand their experiments with agroecology, subsequently creating new markets, an income and greater respect for themselves, and finally standing up for their rights and their desire to further amplify agroecology.
>
> They came out of isolation—in many cases, connected to domestic violence—and into positions of leadership. The key step here was unearthing and organizing the wealth of knowledge of agroecology held collectively by women, which is often diffuse, fragmented and undervalued, even by the women themselves.
>
> *Source*: Galvão Freire (2018)

action, "be it in productive, reproductive, public, or private spaces" (Lopes and Jomalinas 2011; see also: Bezner Kerr et al. 2019).

Gender inequity is particularly oppressive when it intersects with other kinds of inequitable social relations. Figure 8.1, developed by Michel Pimbert and Stefanie Lemke (2018), shows how agroecology can contribute to greater equity when attention is paid to the intersecting balances of power and inequality between actors in the food system, such as farmers and transnational corporations.

To be truly transformative, grassroots organizing, policy advocacy and urban planning in agroecology must involve the leadership of black, Latin and indigenous people, women and other often-marginalized bodies and explicitly work on an equity agenda. The agenda can include, for example,

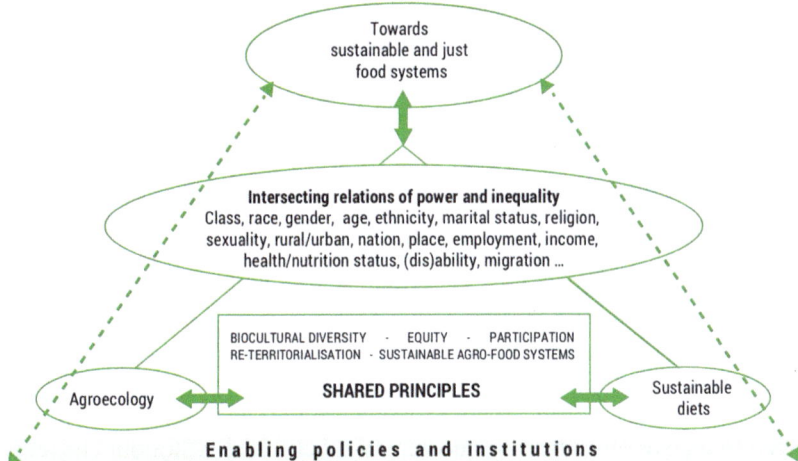

Fig. 8.1 How agroecology and sustainable diets are complementary concepts that can address inequality and contribute to sustainable and just food systems (*Source*: Pimbert and Lemke (2018), concepts based on Rosset and Altieri (2017) for agroecology, Burlingame and Derini (2012) for sustainable diets and Collins (Collins 2000) for intersectionality)

an analysis of inequity that names and addresses racism and discrimination and a deconstruction of structural racism in policies, planning, organizational cultures and education. This can help to topple barriers and provide resources for processes of agroecological transformation led by black and indigenous persons and people of colour (BIPOC). Necessarily, this approach to agroecological transformations would include a process of reparation aimed at undoing historical harm: the inequities and privileges that stem from the theft of indigenous land and livelihoods, plantation agriculture, chattel slavery and the ongoing subjugation of BIPOC cultures and economies around the world.

Farming can offer BIPOC communities the opportunities for economic autonomy while providing safe spaces to gather and celebrate without fear of criminalization or state-based violence (White 2018). As described by the health policy research scholar Ashley Gripper (2020) during the 2020 Black Lives Matter protests,

> Black agriculture provides a way to engage with the disturbing history of this country, that we live in a place built on stolen Indigenous land and the brutal enslavement and stolen labor of my ancestors. It opens the door to us understanding how this all shapes our collective journey toward liberation.

Several Native American communities are also reviving aspects of their traditional agricultural and land-use practices as part of their struggles for self-determination and food sovereignty. As they use traditional seeds to diversify farming, they also seek out, learn, share and affirm the distinct histories of their indigenous communities and unlearn dominant narratives about the supposed inferiority of indigenous food and agriculture (Mihesuah and Hoover 2019). For example, the Anishinaabeg people on the White Earth Indian Reservation in western Minnesota are reviving the cultivation and harvesting of a traditional food of the Ojibwe people, wild rice. As they reclaim their agroecological practices, they honour the legacies of the Ojibwe people to food and farming: an act of resistance to white supremacy and colonial domination (LaDuke 2017).

Similarly, the 40-year-old process of the Women Sanghams, the women's groups which include over 5000 Dalit women who run small farms in the drylands of Telangana, India, highlights the importance of collective reflection and action to overcome domestic and caste-based violence as well as to enhance Dalit women's agroecological pathways to food sovereignty and greater autonomy (The Community Media Trust et al. 2008). Box 8.3 provides another example of a food sovereignty and agroecology that centres anti-caste, indigenous and feminist activism.

Agroecology and anti-hunger research both examine the underlying causes of inequity and point at the need for shifts in governance (Chappell 2018; Wittman et al. 2017). Governance-related measures such as those listed by Alessandra Mora and Pasquale De Muro (2018) can help reinforce such a virtuous cycle between agroecology and equity. They include a greater emphasis on inclusive, people-centred development; better policy monitoring and implementation; decentralization and greater participation of and investment in people who are marginalized and excluded; strengthening the local capacity, accountability and transparency of governments; and stronger implementation of the rule of law. Key too are policy and programmes that address the legacy of racial, ethnic and class inequality to promote equity in food systems.

One effort to do so is the Milan Urban Food Policy Pact of 2015: a commitment by over 160 cities around the world to develop "sustainable food systems that are inclusive, resilient, safe and diverse, that provide healthy and affordable food to all people in a human rights-based framework". In addition, there have been calls for governments to support agroecology by prioritizing implementation of the UN recommendation on the rights of women living in rural areas, adopted in 2016. This covers women's rights to participate and benefit from rural development; to health, education, employment, economic, social and public life, and protection from violence; and to land and other components of ecosystems. Indeed, the role of formal and informal democratic institutions in promoting equity in agroecological transformations cannot be underestimated.

Box 8.3   The Kūdali Intergenerational Learning Centre and the India Food Sovereignty Alliance

Fig. 8.2 Seeds of Resistance event at the Kūdali Intergenerational Learning Centre

(*continued*)

Box 8.3   (continued)

Indian society, like those of many other countries, is highly stratified and structured by class, gender and caste-based inequality. While many organizations work on issues related to agroecology and food sovereignty, the Kūdali Intergenerational Learning Centre and the India Food Sovereignty Alliance embed such work in an anti-caste, anti-patriarchy and anti-capitalist foundation. It is from this starting point that they engage with the politics of food sovereignty and agroecology.

Kūdali (which means joining, meeting) is a physical learning centre in Telangana, India, and a transformative space for intergenerational, intercultural learning and popular education initiatives. Kūdali is a part of the food sovereignty movement led by Dalits, the indigenous Adivasis, and small peasants, pastoralists and other 'people who eat' (sometimes referred to as consumers). Kūdali supports the indigenous philosophy also promoted by the *Buen Vivir* thinking in Latin America, in which food sovereignty and social justice are a critical framework of action and practice.

Members of India's Food Sovereignty Alliance know that reaching out to rural and urban children and youth in schools and universities for a dialogue on food sovereignty is critical for the future of this movement. Interactions with youth take place in schools and universities as well as in the movement's learning centre.

At Kūdali, critical thinking on collective futures is encouraged through meeting agroecological farmers; visiting their fields and eating the food grown on their farms: understanding the links between people, the ecology, culture and food and questioning the actors and structures that block food sovereignty; learning to work with soil, dung and seeds; and expressing their views in diverse creative ways including art, song and theatre.

Together, by affirming their culture, rights and agency, the India Food Sovereignty Alliance directly confronts inequity as the basis of realizing the promise of food sovereignty.

Source: Yakshi (www.yakshi.org.in) and the India Food Sovereignty Alliance (https://foodsovereigntyalliance.wordpress.com/)

Disabling Conditions

Inequity is, as we have shown, pervasive across all social systems. Women and people from lower castes, minority ethnic groups and races often find that the rules of the game are heavily biased against them in society because they have been historically structured around the physical needs, capabilities and political interests of the powerful, who designed them in the first place (Goetz 1997).

The food system has a long history of dispossession and exploitation of people of colour. Today, they must navigate structurally racist societies with longstanding patterns of inequity, reverberations of historical trauma, anti-blackness and white supremacy that play out in food systems as in society. The study of the relationship between racism and food systems has been most thorough in the United States, with many accounts of how black peoples, indigenous peoples and people of colour have faced, first, exclusion over ownership and, in the case of indigenous peoples, theft of land. Today, ongoing structurally discriminatory policies continue to raise barriers for BIPOC people to access land, financing, education and other resources that are fundamental to building agroecology (see Chap. 4 on access to nature).

On the other hand, many of the emerging 'alternative food networks' and initiatives related to agroecology, farmers' markets and community-supported agriculture are driven by privileged interests and steeped in white culture and values—or what is often referred to as whiteness. These initiatives are often exclusionary and perpetuate harm (Slocum 2007). Racism is closely tied to the injustices causing poverty, hunger and malnutrition (Holt-Gimenez and Harper 2016). Ensuring equity of access to healthy food, resources and dignified, living-wage jobs would make meaningful contributions towards more justice and equity in the food system. In addition, Eric Holt-Gimenez and Breeze Harper (2016) note, resources that are being used in methods for healing historical trauma, and working through immobilizing feelings of internalized oppression, fear, hopelessness and guilt, can be brought into the agroecology movement.

Women are particularly hard-hit by these persistent burdens. The dominant agricultural development model is "largely gender blind, patriarchal, and indifferent to human rights, including women's rights", note Ana Paula Lopes and Emilia Jomalinas (2011) as it ignores and undermines the important knowledge and perspectives of women and other systematically marginalized people in agriculture and rural communities. Around the world women are still largely responsible for collecting water, fuel, cooking, caretaking and agricultural tasks, yet continue to have less access and rights to a variety of resources, health services, and care and decision-making.

An ongoing policy focus on commercialized, large-scale, export-oriented agriculture, for instance, has in some cases led men to migrate to urban areas to find work, increasing the pressure on women to care for their families' health and food security (Deepak 2013). For example, strong correlations have been found between the use of agrochemical inputs and gender inequity in the Sahel; here, men are the main recipients of state-subsidized chemical fertilizer (Brescia 2017) while women generally do not have access to them. This has left women to be the first to experiment with soil fertility-enhancing practices grounded in agroecological principles. While this is a positive development, recognition and support for these women's efforts and knowledge has not often followed.

Similarly, 'green revolution'-type approaches have maintained or exacerbated disadvantages for poor and women farmers and other marginalized groups (Negin et al. 2009). In numerous countries, women have been effectively blocked from engaging in agroecological innovation through means such as violence—from the psychological to the physical. Policy blindness to these inequities maintains conditions of inequality and patriarchy (Schwendler and Thompson 2017) in communities, and beyond.

Intersecting inequities in rurality, peasant status, caste, race, class, religion, health, age and gender, along with aggressive large-scale land grabs and rising food prices, have in many places weakened the position of rural women even further. Such persistent inequity can disable the self-organizing processes in communities that drive agroecology transformations. For example, Rachel Bezner Kerr et al. (2019) found that in Malawi the intersection between gender and class dynamics, combined with state policies, undermines agroecology-related processes that hold the promise of addressing inequality. Similarly, Manoj Misra (2017) argues that any strategy to address rural malnutrition in Bangladesh through agroecology must first resolve the existing conflicts between opposing agricultural classes, such as between landholders and workers without land.

Despite such issues, within agricultural research and extension, as well as in ministries for environment and development, units specifically set up to help integrate gender or indigenous peoples' issues in different departments have been notoriously under-resourced in staff and funds—and marginalized (Goetz 1997). Equity-sensitive policy proposals are rarely reflected in budgetary allocations. This de-linking of progressive policy statements from actual budgetary re-orientations and commitments often occurs in the public expenditure planning process, thus effectively excluding gender and other intersecting equity issues in national or local agricultural development and land-use planning. The under-representation of women, minority ethnic groups and BIPOC people is also an obstacle to the institutionalization of equitable practices in universities, government departments and society (Sian 2019).

The obstacles to equity and inclusion are also embedded in the operational procedures and service delivery of development organizations. In project design and implementation, relatively little attention is usually paid to the unequal division of labour, power and resources between women and men, as well as between groups differing in regard to class, age, race, ability, sexuality and ethnicity. The interventions of bureaucracies such as research institutions or government ministries of agriculture and rural development have often generated and exacerbated intersectional inequities, ultimately harming people and livelihoods (Goetz 1997).

REFERENCES

Bezner Kerr, R., Hickey, C., Lupafya, E., & Dakishoni, L. (2019). Repairing Rifts or Reproducing Inequalities? Agroecology, Food Sovereignty, and Gender Justice in Malawi. *The Journal of Peasant Studies, 46*(7), 1499–1518.

Brescia, S. (Ed.). (2017). *Fertile Ground: Scaling Agroecology from the Ground Up.* Oakland: Food First.

Burlingame, B. (2012). Preface. In B. Burlingame & S. Dernini (Eds.), *Sustainable Diets and Biodiversity: Directions and Solutions for Policy, Research and Action* (pp. 6–8). Rome: FAO.

Chappell, M. J. (2018). *Beginning to End Hunger: Food and the Environment in Belo Horizonte, Brazil, and Beyond.* Oakland: University of California Press.

Collins, P. H. (2000). *Black Feminist Thought: Knowledge, Consciousness, and the Politics of Empowerment.* London and New York: Routledge.

de Marco Larrauri, O., Pérez Neira, D., & Soler Montiel, M. (2016). Indicators for the Analysis of Peasant Women's Equity and Empowerment Situations in a Sustainability Framework: A Case Study of Cacao Production in Ecuador. *Sustainability, 8*(12), 1231.

De Schutter, O., & Campeau, C. (2018). Equity, Equality and Non-discrimination to Guide Food-System Reform. *UNSCN-News: Advancing Equity, Equality and Non-discrimination in Food Systems: Pathways to Reform, 45*, 7–14.

Deepak, A. C. (2013). A Postcolonial Feminist Social Work Perspective on Global Food Insecurity. *Affilia, 29*(2), 153–164.

FAO. (2018). *Catalysing Dialogue and Cooperation to Scale Up Agroecology: Outcomes of the FAO Regional Seminars on Agroecology*. Rome: FAO.

Galvão Freire, A. (2018). Women in Brazil Build Autonomy with Agroecology. *Farming Matters, 34*(1), 22–25.

Goetz, A. M. (1997). *Getting Institutions Right for Women in Development*. Zed Books.

Gómez, L. F., Ríos-Osorio, L., & Eschenhagen, M. L. (2013). Agroecology Publications and Coloniality of Knowledge. *Agronomy for Sustainable Development, 33*(2), 355–362.

Grey, S., & Patel, R. (2014). Food Sovereignty as Decolonization: Some Contributions from Indigenous Movements to Food System and Development Politics. *Agriculture and Human Values, 32*(3), 431–444.

Gripper, A. (2020). We Don't Farm Because It's Trendy; We Farm as Resistance, for Healing and Sovereignty. *Environmental Health News*.

Holt-Gimenez, E., & Harper, B. (2016). *Food—Systems—Racism: From Mistreatment to Transformation*. Food First.

Khadse, A. (2017). Women, Agroecology & Gender Equality. *Focus on the Global South, India*. Retrieved from https://focusweb.org/system/files/women_agroecology_gender_equality.pdf.

La Via Campesina. (2016). Gender Diversity in the Peasant Movement. Accessed on 28 Nov 2020 at https://viacampesina.org/en/gender-diversity-in-the-peasantmovement/.

LaDuke, W. (2017). *All Our Relations: Native Struggles for Land and Life*. Haymarket Books.

Lambrecht, I. B. (2016). "As a Husband I Will Love, Lead, and Provide." Gendered Access to Land in Ghana. *World Development, 88*, 188–200.

Lopes, A. P., & Jomalinas, E. (2011). *Agroecology: Exploring Opportunities from Women's Empowerment Based on Experiences from Brazil. Feminist Perspectives Towards Transforming Economic Power*. Toronto, Mexico City, Cape Town: Association of Women's Rights in Development.

Mihesuah, D., & Hoover, E. (2019). *Indigenous Food Sovereignty in the United States: Restoring Cultural Knowledge, Protecting Environments, and Regaining Health* (Vol. 18). University of Oklahoma Press.

Misra, M. (2017). Moving Away from Technocratic Framing: Agroecology and Food Sovereignty as Possible Alternatives to Alleviate Rural Malnutrition in Bangladesh. *Agriculture and Human Values, 35*, 473–487.

Mora, A., & De Muro, P. (2018). Inequality and Malnutrition. *UNSCN-News: Advancing Equity, Equality and Non-discrimination in Food Systems: Pathways to Reform, 45*, 15–24.

Negin, J., Remans, R., Karuti, S., & Fanzo, J. C. (2009). Integrating a Broader Notion of Food Security and Gender Empowerment into the African Green Revolution. *Food Security, 1*(3), 351–360.

Pimbert, M. P., & Lemke, S. (2018). Food Environments: Using Agroecology to Enhance Dietary Diversity. *UNSCN News, 43,* 33–42.

Rosset, P. M., & Altieri, M. A. (2017). *Agroecology: Science and Politics.* Winnipeg: Fernwood.

Schwendler, S. F., & Thompson, L. A. (2017). An Education in Gender and Agroecology in Brazil's Landless Rural Workers' Movement. *Gender and Education, 29*(1), 100–114.

Sian, K. P. (2019). *Navigating Institutional Racism in British Universities.* Palgrave Macmillan.

Slocum, R. (2007). Whiteness, Space and Alternative Food Practice. *Geoforum, 38*(3), 520–533.

Soler, M., Ferre, M. R., & Roces, I. G. (2019). Feminist Agroecology for Food Sovereignty. *Cultivate!*

The Community Media Trust, Satheesh, P. V., & Pimbert, M. P. (2008). *Affirming Life and Diversity: Rural Images and Voices on Food Sovereignty in South India.* London: IIED.

van der Ploeg, J. D. (2018). *The New Peasantries: Rural Development in Times of Globalization.* Earthscan Food and Agriculture.

White, M. M. (2018). *Freedom Farmers: Agricultural Resistance and the Black Freedom Movement.* UNC Press Books.

Wittman, H., Chappell, M. J., Abson, D. J., Kerr, R. B., Blesh, J., Hanspach, J., et al. (2017). A Social–Ecological Perspective on Harmonizing Food Security and Biodiversity Conservation. *Regional Environmental Change, 17*(5), 1291–1301.

Wretched of the Earth. (2019). *An Open Letter to Extinction Rebellion.*

CHAPTER 9

Domain F: Discourse

Abstract In this chapter, we examine how discourse—or the ways in which language is used to frame debates, policy and action—is a critical domain for agroecology transformations. A range of different types of actors (e.g. politicians, private companies, activists) use a process called 'framing' to convey their interpretation of agroecology where they 'simplify and condense' its complexity to align with their own views and ideologies. We present seven main frames across a spectrum from those that tend to disable a transformative agroecology (e.g. 'feed the world') to those that are most likely to enable political agroecology (e.g. 'food sovereignty'). Notably all of these frames are at times being deployed in both productivist and depoliticized (regime-reinforcing) ways and also as a part of a transformative politics of political agroecology at different times by different actors.

Keywords Discourse • Framing • Food sovereignty • Right to food • Definition

Discourse—the ways in which language is used to frame debates, policy and action—is a critical domain in shaping agroecology transformations (Mier y Terán Giménez Cacho et al. 2018). A range of actors with different social status and worldviews engage in debate over the agroecological "terrain of ideas, of theoretical constructs" (Giraldo and Rosset 2018).

C. R. Anderson et al., *Agroecology Now!*,
https://doi.org/10.1007/978-3-030-61315-0_9

This carries implications for which pathways for food system transformation are considered socially legitimate and high priority (Dryzek 2013) and thus how they are resourced and supported. Discourse "directly shapes and conditions the policies and actions taken" (Ajates Gonzalez et al. 2018; Lamine 2017; Loconto and Fouilleux 2019; Pimbert 2015), not just the goals, metrics, standards and practices implied when discussing agroecology.

The discourse on agroecology is shaped by producers' organizations and other civil society groups, governments, the Food and Agriculture Organization of the United Nations (FAO) and other multilateral institutions, researchers, media and the private sector. It is thus not surprising that definitions of agroecology and its role vary hugely, even though it is ostensibly a concept and practice that unifies a diversity of actors (Loconto and Fouilleux 2019; Pimbert 2018a; Rivera Ferre 2018). Despite efforts to advance one particular meaning over another, agroecology is malleable and subject to political processes.

This entails a discursive process, 'framing', used to interpret agroecology in a way that simplifies its complexity and emphasizes characteristics that align with a specific agenda (Benford and Snow 2000). By selectively drawing on and interpreting agroecology through the lens of their own cultural values, beliefs and ideologies, particular actors can frame it in ways useful to them (Geels and Verhees 2011; Steinberg 1998). Below we elaborate on seven key discursive frames we have identified as underpinning key debates on agroecology.

Our analysis suggests that these frames each have a different underlying political basis and intention; these render them more or less enabling or disabling of a transformative agroecology. In Fig. 9.1 you will see a spectrum of frames. At the red end are those that tend to disable a transformative agroecology. At the green are those emphasizing the agency of communities and food producers, as well as resonance with local cultures and are supportive of what we have outlined above as a political agroecology. In the middle are the rest: frames that are much more ambiguous in use and potential. Notably all of these frames are at times being deployed in both productivist and depoliticized (regime-reinforcing) ways and also as a part of a transformative politics of political agroecology at different times by different actors. In the following section we discuss each of these seven frames.

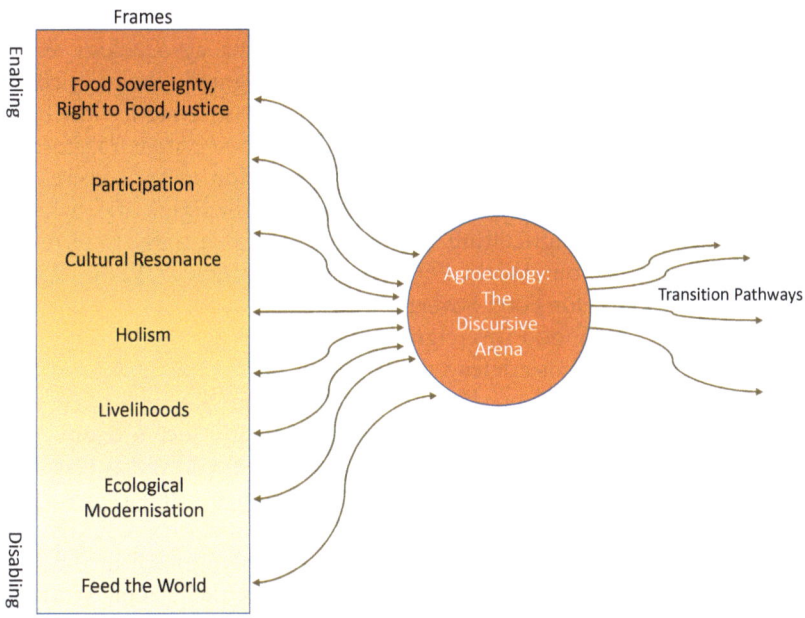

Fig. 9.1 Discourse around agroecology is shaped by different frames that can have both enabling and disabling effects on political agroecology. Some frames, towards the top end of the figure, are much more enabling, while the frames towards the bottom are more likely to have a disabling effect. Other frames, in the middle, are more ambiguous

FOOD SOVEREIGNTY, THE RIGHT TO FOOD AND JUSTICE

To advance a transformative agroecology, actors frequently use three related frames that present an enabling set of values, beliefs, principles and worldviews: food sovereignty, human rights and justice. These three inter-meshed frames have been developed dialectically along with terms and concepts such as 'transformative agroecology' (Méndez et al. 2015; Levidow 2015, #360), 'political agroecology' (Méndez et al. 2015) and 'radical, movement-based agroecology' (Holt-Giménez and Altieri 2013). They are rooted in a shared position: that profound political and systemic change is needed to address power relations and advance equity and democracy in the food system.

Many social movements and scholars frame agroecology as an insepa-rable component of, and pathway towards, *food sovereignty* (Nyeleni 2015;

World Forum of Fisher Peoples (WFFP) 2017). While political activities such as protest and advocacy figure large in this arena, agroecology represents an on-the-ground articulation of food sovereignty in the practices of food producers. Specifically, the concept of food sovereignty has been taken up around the world as a political project of food system transformation, rooted in agroecology and the democratization of agriculture and food. As such, it embodies a discourse that affirms the rights of peoples to define their food and agriculture systems as well as their rights to territory and self-determination (World Forum of Fisher Peoples (WFFP) 2017; Nyeleni Movement for Food Sovereignty 2007).

Drawing on both food sovereignty and *human rights* frameworks, civil society organizations also defend the rights of peoples to healthy and culturally appropriate food produced through ecologically sound and sustainable methods—as well as the rights of food producers to use and manage lands, territories, waters, seeds, livestock and biodiversity (Nyeleni 2015). Thanks to the leadership of social movements and civil society organizations such as La Via Campesina, FIAN International, GRAIN and CETIM, many of these rights are now officially recognized in the UN (2018) Declaration on the Rights of Peasants and Other People Working in Rural Areas (UNDROP). UNDROP radically reframes the dominant legal paradigm by introducing new individual and collective rights to nature and food sovereignty that go beyond the 'right to food' (Claeys 2015). Governments too have been using the human rights frame for agroecology, as it is embedded, for example, in the discourse of New Latin American or New Andean Constitutionalism. These refer to a wave of constitutional reform in Ecuador, Bolivia and Brazil that aim to enhance participatory democracy and recognize the rights of marginalized groups such as indigenous communities.

Finally, civil society organizations and researchers also frame agroecology by associating it with radical forms of *justice* that challenge the dominant food regime. This frame is sometimes combined with food sovereignty. Researchers, for example, advocate for 'distributive and procedural justice' in relation to agroecology, inquiring respectively into who gets access to what resources as well as who makes decisions about resources, and how (Chris Maughan et al. 2020; Schwendler and Thompson 2017). Further, Michel Pimbert (2018b, p. 31) calls for 'cognitive justice' that recognizes "the right of different forms of knowledge and their associated practices, livelihoods and socio-ecological contexts to coexist" (also see Chap. 5 on the knowledge domain). Introducing yet another aspect Cristian

Timmermann and Georges Félix (2015) examine the ways in which agroecology enables 'contributive justice'. The latter refers to an agroecological work environment where people can have the opportunity to develop skills and be creative and productive while paying attention to a fairer distribution of meaningful work and tedious tasks. Civil society organizations link agroecology with 'climate justice' (e.g. Friends of the Earth International 2015; La Via Campesina 2018). Finally, movements and researchers are increasingly pointing at the link between agroecology and gender justice, or even a 'feminist agroecology' (Articulação Nacional de Agroecología (ANA) 2018; Bezner Kerr et al. 2019). Some are emphasizing approaches that simultaneously challenge colonialism, racism, capitalism and patriarchy in the food system.

While these frameworks are rarely used in a disabling way, in some cases a depoliticized and sanitized version of these terms is deployed by groups or individuals. The governments of France, Ecuador and Venezuela, for example, have also used the food sovereignty frame in relation to agroecology but have interpreted it narrowly as national or regional food self-sufficiency. This can feed into nationalistic, exclusionary tendencies or become a way to promote national corporate interests. Moreover, the private sector sometimes adopts a rights-based discursive frame when placing intellectual property rights on seeds (see Chap. 4 on rights and access to nature), which runs counter to transformative agroecology.

Finally, many actors use the notion of 'rights' in the context of the neoliberal refrain regarding the right to choose which products or technologies to use. For example, farmers are seen as individual consumers who should have the unimpeded right to use industrial chemicals or consumers should be free to choose the products they like, without acknowledging the multifarious constraints and factors going into such 'choices'. An approach focusing on individual choice obscures all the power dynamics that limit the options available to farmers or citizens.

PARTICIPATION

Participation and democratization are at the heart of transformative agroecology, implying that the central agency lies with organizations of agricultural producers and citizens. The frame of participation provides a vision and a basis for the process and governance-oriented principles of agroecology (HLPE 2019).

For example, the International Panel of Experts on Sustainable Food Systems (IPES-Food 2019) points out that by shifting the focus from agriculture to the entire food system, a wider range of stakeholders can be meaningfully involved in designing and assessing policies for agroecological transformation, thereby linking participation with the holism frame, which emphasizes the interconnectedness of elements in the agroecological food system (see below). Moreover, several authors argue that promoting multi-actor collaborations at the territorial scale, for instance in the form of food policy councils, is a particularly enabling factor in agroecological transformations (Lamine et al. 2019).

This frame has also been shaped by experiences with participatory governance in agroecology. A widely cited institutional example took place in Brazil, where the official integration of agroecology into public policy and discourse was shaped by a long history of interactions between the state, social movements, agricultural producers and researchers (Schmitt et al. 2017). This social dialogue played a key role, both in building convergence within civil society around a shared framing of agroecology and in proactively shaping the state's understanding that the development of agroecology requires a state-civil society dialogue. It led to the adoption of the National Policy on Agroecology and Organic Production (PNAPO) and the associated plans guiding its implementation.

FAO too considers participatory, 'responsible', governance as key to agroecological transitions (FAO 2018a). It argues that transparent, accountable and inclusive governance is required at multiple scales, for example to ensure equitable access to nature, including land. Further discursive links can be made between participation and other domains of agroecology transformation. Some associate agroecology with the commons, stressing collective approaches to environmental stewardship and knowledge (Nyeleni 2015; see also Chap. 4). Pimbert (2018a) calls for different forms of radical democracy and active citizenship in the governance of research and knowledge production for agroecology. Others emphasize the collaborative character of agroecological systems of exchange, embodied, for example, by cooperatives, participatory guarantee schemes and community-supported agriculture, which are often community-based, embrace participatory decision-making and strive towards inclusivity.

Although the participatory framing of agroecology generally enables transformation, the extent to which participation and participatory democracy are realized in practice, as part of an agroecological transition, is uneven. In this regard, Arnstein's ladder of citizens' participation is a useful reminder that participation can range from manipulation to more empowered forms in which people have control (Arnstein 1969). The field of participatory development and public participation in policy-making has long been characterized by narrow and perverted approaches to participation that are not guided by participants but rather by narrow agency within a pre-determined framework and are often used to justify and advance already existing agendas of governments, planners or NGOs (Cooke and Kothari 2001). As Raquel Ajates Gonzalez et al. (2018) point out, in the development of France's national strategy on agroecology, participation of civil society appeared to be limited to consultations on policy proposals, with limited influence on the final policy outcome, and an evaluative role along or at the end of the implementation process.

Further illustrating the co-optable nature of the participation frame, the World Economic Forum (WEF) and the Food Action Alliance that are advocating for the so-called Fourth Industrial Revolution (4IR) are also calling for 'transformative partnerships' and the need to create more sustainable and inclusive food systems. In this regard, 4IR actors claim to enable women entrepreneurs, youth and small farmers, particularly in Africa, to access 4IR technologies and new markets. Specifically, their inclusion in the world economy is to take place through digital platforms for food value chains (e.g. Technical Centre for Agricultural and Rural Cooperation (CTA) 2019). These platforms are virtual marketplaces that match supply and demand across the globe for agricultural inputs, equipment, products and services. In practice, therefore, the main vehicles for 4IR food system transformation are market-driven solutions led by the private sector and facilitated by the state, without meaningful spaces and means for other actors to participate in decision-making on this transformation. These conceptions of participation reinforce asymmetric power relations in the dominant regime. They are incompatible with political agroecology.

Cultural Resonance

Framing agroecology as a culturally appropriate, place-based form of agriculture and food provisioning enables transformation. The agroecological organic coffee movement in Chiapas and the Mesoamerican Campesino a Campesino (CaC) network, for example, are inspired by cultural frames linking liberation theology, values of autonomy, love for Mother Earth, defence of territory and culture and the cosmovisions of Mesoamerican peoples (Mier y Terán Giménez Cacho et al. 2018). In the United States, black farmers' collectives are promoting 'Afro-ecology' as a form of agroecology shaped by Afro-indigenous life experience and traditions (Black Dirt Farm Collective 2016).

'Biocultural diversity' is another concept that researchers and communities use to describe agroecology's cultural embeddedness—the interrelatedness of biological and cultural diversity in territories (Pimbert and Borrini-Feyerabend 2019). In policy supporting agroecology, an example of cultural resonance is the 'New Andean Constitutionalism' in Ecuador and Bolivia. This approach embodies 'epistemologies of the South' (Santos 2015)—indigenous cosmovisions and knowledge systems—as the basis for governance, food sovereignty and agroecology (Schilling-Vacaflor 2011).

Beyond production, some groups are calling for 'culturally appropriate diets' as part of agroecology (Baker et al. 2019; FAO 2018a) and, more broadly, for culturally diverse definitions of a 'good life'. They include movements for Ecological Swaraj in India, Eco-Ubuntu in South Africa and Buen Vivir in Latin America.

Although, as a frame, cultural resonance is largely enabling, some deploy it in ways that undermine political agroecology. Some practitioners of Zero Budget Natural Farming in India—which has been celebrated as an agroecological innovation—have adopted a Hindu nationalist stance, in which religious and ethnic minorities and people marked as lower caste are viewed as inferior (Bhattacharya 2017; Khadse et al. 2017). Cultural discourse that is prejudiced and racist clearly violates the principles of equity that underpin agroecology (see Chap. 8 on equity). In another case, working within the dominant regime, the Alliance for a Green Revolution in Africa (AGRA) declares on its website that "African farmers need uniquely African solutions" to sustainably increase their productivity and access markets. Here, the rhetoric of cultural resonance (African pride) obscures the fact that the AGRA website primarily advances Western technologies and corporate interests in Africa.

These examples clearly show how the notion of cultural resonance can also be deployed in superficial, disingenuous ways that go against the heart of agroecology, by strengthening xenophobic sentiments or advancing industrial agriculture.

HOLISM

Agroecological transformation can be enabled when agroecology is incorporated into wider calls for holism—a frame emphasizing the interconnectedness of elements in the agroecological system. Holism reflects a significant break from sectoral thinking, which suggests that agroecology is solely about agriculture and only concerns farmers, and contrasts with reductionist thinking and compartmentalization. A holistic framing chimes with agroecology's embrace of complexity and interconnectedness as a way of triggering wider social transformation within the agri-food system.

In the Declaration of the International Forum on Agroecology (2015), social movements conceptualize agroecological transitions as cutting across multiple agricultural sectors and bridging political, economic and cultural dimensions of food systems. For researchers, this perspective implies adopting a transdisciplinary, participatory, action-oriented approach to investigation, combining the natural and social sciences with the local knowledge of practitioners and consumers (Méndez et al. 2015; see also Chap. 5). Linking the intersectoral nature of agroecology with the need to integrate related knowledge systems, researchers and social movements often emphasize that agroecology is simultaneously "a movement, a science, and a practice" (Wezel et al. 2009).

Systems, rather than sectoral, thinking has also been an important variant of this frame. FAO, for instance, calls for a systems vision on agricultural policy development that "maximizes synergies within the food system, mitigates negative externalities and minimizes harmful competition between agricultural sectors" as well as between agriculture and other sectors (FAO 2014, 2018c). In the same spirit, the International Panel of Experts on Sustainable Food Systems (IPES-Food 2019) advocates for an umbrella strategy for food system transformation in Europe that integrates policy areas currently handled by separate directorate generals and committees.

From a critical perspective, not all claims to holism lend themselves to a transformative agroecology; indeed, they may play a role in co-opting its radical potential. Several studies note that the governments of France (Ajates Gonzalez et al. 2018) and China (Shiming and Gliessman 2017) have both engaged in discourse and created policies that gesture towards holism. These have, however, been strongly shaped by a reductionist scientific and technical understanding of agroecology that aligns well with the dominant regime but lacks reference to intersectoral linkages and the socio-political aspects of agroecology. Further weakening the transforma-tive potential of holism, proponents of the 4IR in food and agriculture are also calling for a 'system-wide' transformation based on the alignment of multiple actors for deploying 4IR technologies across the globe and enabling integrated value chain investments such as the Grow Africa pro-gramme (World Economic Forum (WEF) 2018). Although Grow Africa aims to facilitate cross-sector policy dialogue with the government, private sector, research and civil society, its ultimate aim is to transform African agriculture through private sector investment in specific agricultural com-modity value chains (e.g. cassava in Nigeria, mango in Burkina Faso) and connect African farmers with national, regional and international markets. This approach puts the private sector in a privileged position and favours a market-led rather than a holistic agricultural transformation.

LIVELIHOODS

As a frame, livelihoods can either enable or disable a transformative agro-ecology. From an enabling perspective, this frame reveals how agroecol-ogy can strengthen the livelihoods and well-being of smallholder food producers, indigenous peoples, women and young people and how they in turn are well suited to advance agroecology. It also emphasizes agroecol-ogy's connections with the agency and autonomy of food producers, fam-ily farming as a way of life and the centrality of rural people's livelihoods (IPES-Food 2016; van Walsum et al. 2014). Agroecology is thus sharply contrasted with the dominant regime's dehumanizing, modernizing, urbanizing, capitalist logics in agriculture, where livelihoods are an exter-nality or an indirect effect.

A wide range of actors deploy the livelihood frame to enable agroecol-ogy. Food sovereignty movements have long argued for the importance of farmers' agency and livelihoods in agroecology (Nyeleni 2015). La Via Campesina, for example, has emphasized the value of the peasant way of life (Desmarais 2008), invoked too in their framings of agroecology. Social movements and farming families often highlight how agroecology can improve farmers' livelihoods by helping them rely less on, or avoid, input and credit markets, expensive technologies and exploitative long supply chains (Rosset and Martinez-Torres 2012). Similarly, FAO recognizes that the multifunctionality of family farmers allows them to act holistically on multiple dimensions of agroecology. Their multiple functions include pro-ducing most of the world's food, acting as stewards of nature by preserv-ing and developing biodiversity, preserving and sharing traditional knowledge, and contributing to the resilience of people and nature. Importantly, when empowered, they help strengthen the economic viability of rural areas (FAO 2018b).

This frame is also closely linked to knowledge, creativity and solidarity economy, an ethical and values-based approach to economic well-being that prioritizes the welfare of people and planet over profits and economic growth. Through this, it reveals how agroecology plays an important role in creating meaningful employment as well as fair livelihoods for food producers (FAO 2018a; Timmermann and Félix 2015; World Forum of Fisher Peoples (WFFP) 2017; see Chap. 6 on the systems of economic exchange domain). Relatedly, an emerging line of discourse associates agroecology with alternative definitions of well-being that include fair livelihoods such as de-growth and *Buen Vivir* (Kothari et al. 2015).

Prior to the 2018 election of Brazilian President Jair Bolsonaro, the institutional space for agroecology in the country had been opened up by the formal recognition of family farming as an economically viable form of agriculture—one that increased its social legitimacy and public visibility while contributing to the emergence of a discourse that strongly associated a political agroecology with family farmers (Lamine 2017). In addition, a recent empirical study of experiences across Europe emphasizes the economic potential of agroecology for sustaining livelihoods of family farmers (van der Ploeg et al. 2019).

The livelihoods frame, however, has also been disabling for a transformative agroecology. First, proponents of the corporate-led 4IR also claim to be opening up new employment opportunities and new markets for small farmers, particularly in Africa, by creating an inclusive digital environment (e.g. Technical Centre for Agricultural and Rural Cooperation (CTA) 2019). In this model, 'livelihood' is often reduced merely to income or economic returns—assuming that access to new markets will increase income and therefore increase farmers' ability to achieve food security by buying food—what Jahi Chappell (2018) has argued is a potential form of 'neo-productivism'. This approach, however, is incompatible with political agroecology on many levels. It creates dependency on expensive inputs, disenfranchises food producers as agents of change and focuses on the production of monocrops for global markets instead of the development of regional food systems, food producer agency and diverse, healthy diets.

Similarly, the livelihoods frame is also often flipped so that small-scale, family-based agriculture is trivialized in favour of a business-focused framing of livelihoods. This can demobilize food producers and rural communities interested in agroecology by preventing them from launching or expanding agroecological experiments. Such discursive frames often label peasants, traditional rural communities and traditional forms of agriculture as poor, backward, low quality, inefficient or unproductive, suggesting that agriculture is inherently a form of drudgery (Isgren 2016; Schneider 2015). At the same time, they may present large-scale producers and industrial forms of agriculture as modern, productive, tidy, entrepreneurial and representative of 'good' farming and insist that it is in farmers' and society's best interests to minimize the number of people unfortunate enough to be farmers.

In contemporary China, negative discourse on peasants (*nongmin*) and small-scale agriculture has justified and shaped agricultural policies that aim to reduce the number of peasants and promote agricultural modernization and urbanization through novel forms of industrialization (Schneider 2015; Si et al. 2018). Although claiming to enable small farmers' livelihoods, proponents of 4IR also aim, for example, to foster "a new breed of young ICT 'agripreneurs'" (Technical Centre for Agricultural and Rural Cooperation (CTA) 2019, p. 10). This kind of framing attempts to minimize the assumed hardship and drudgery of farming by industrializing it and minimizing the number of people 'subjected' to it—rather than seeking to decrease the marginalization, monoculturalization, low pay and low respect often afforded to family farmers.

Box 9.1 A spotlight on the problematic nature of the "Innovation Frame" for political agroecology
The framing of agroecology as a sustainable or green innovation has taken hold over the past few years. While embraced by a wide range of actors, this frame can undermine the potential of a transformative agroecology for the following reasons:

(1) The innovation frame often reduces agroecology to its technical dimensions. It positions agroecology as one of multiple innovations in a wider toolbox containing purely technological approaches, rather than viewing it as an alternative paradigm and political transformation of the food system.
(2) Innovation is deeply tied to capitalist and neoliberal logics of economic development and productivism. Indicators of whether

(*continued*)

Box 9.1 (continued)

agroecology is innovative tend to be based on narrow productivity and profitability measures on individual farms and of individual crops. This marginalizes or erases all of the multiple functions (see Chapter 2, page 18) of agroecology, which are also qualitative, social and political.

(3) Discursively, innovation is often directly tied to modern technology and thus agroecology is often viewed as backward in this context.

In their recent report *Agroecology and Other Innovations* (2019), the High Level Panel of Experts on Food Security and Nutrition essentially argued for a demotion of the innovation frame, claiming instead that the value of "agroecology and other innovations" need to be assessed for their capacity to realize people's agency and rights. Within this significant global food policy process, proponents of a transformative agroecology were able to assert their power in reframing the debate and produce a high-profile UN report that centres a transformative framing of agroecology.

Source: Maughn and Anderson, Forthcoming Publication.

ECOLOGICAL MODERNIZATION

Recognizing the ecological imperative to address the multiple crises in the food system, discursive frames that emphasize ecological modernization (EM) have gained international traction. Many calling for it are advocates of high-tech approaches to food system transformation, such as sustainable intensification, climate-smart agriculture and the 4IR (Pimbert 2015). As an approach to environmental policy-making that supports the dominant food system, EM describes an ecological restructuring of the capitalist political economy and the associated industrial food system (Dryzek 2013; Horlings and Marsden 2011).

Most who promote EM are in the private sector, science, government and multilateral organizations, and they perceive environmental degradation—caused in part by polluting, resource-intensive food and agriculture systems—as an impediment to continued, albeit greener, economic growth. This framing informs the European Commission's Bioeconomy Strategy, which aims to support "the modernisation and strengthening of the EU industrial base through the creation of new value chains and greener, more cost-effective industrial processes" (European Commission 2018; Levidow 2015).

EM is particularly disabling in the agroecology context because it appears to contribute to many of the immediate goals of the environmental movement—such as reducing pesticide use and increasing energy efficiency and the availability of mass-produced 'sustainable' food—through mostly technological solutions in large-scale agricultural systems. But such approaches do nothing to address the systemic, political and social underpinnings of the current crises. Through the EM frame, agroecology becomes pigeonholed as one small subset of a broader range of sustainable food system practices, rather than a transformative, even subversive, paradigm.

In France, for instance, government discourse has framed agroecology as an essentially economic rather than environmental policy and presented the environmental performance of farms, achieved through increased resource efficiency and reduced use of chemical inputs, as a lever for raising productivity and competitiveness and for generating further economic benefits. Similarly, in China the government has emphasized how 'ecological civilization' enables eco-agriculture, reflecting the EM-inspired view that environmental sustainability and economic growth can be reconciled (Loconto and Fouilleux 2019). As is the case in other countries, the Chinese government approaches citizens as potential consumers of green products and services, rather than as political agents of change.

Lummina Horlings and Terry Marsden (2011) observe that in past decades, the dominant food regime has privileged the pathway of a 'weak' EM frame, focusing on technological solutions for the sustainable use of natural resources. Sustainable intensification and the 4IR reflect this trend. Focusing on improving food availability and stability—and in line with the 'feed the world' frame (see below) as well as 'weak' EM—the discourse on sustainable intensification promotes emerging technological innovations (such as next-generation biotechnologies, robots and blockchain) to increase productivity 'sustainably' (Bernard and Lux 2016; HLPE 2019; see also Box 9.1 on innovation) and international trade.

For instance, FAO (2017, 2019), the Global Forum for the Future of Agriculture (2020) and the WEF (which drives the 4IR) are using the EM frame to promote a market-driven, science-led food system transformation (World Economic Forum (WEF) 2018). The technologies promoted and the focus on international trade are disabling factors for agroecology. Another weakness in the EM frame is a view of nature as "supplier of resources", "a recycler of pollutants" and an enabler of convivial green lifestyles (Dryzek 2013, p. 170). Indigenous approaches to agroecology, as well as emerging research on agroecology and economic de-growth, also question the 'green growth' model of EM.

While countries practising stronger EM approaches have been opening up environmental policy-making to a wider range of actors, including green groups, John Dryzek (2013) concludes that this privilege is often limited to already empowered actors and 'reformist environmental groups' (see also the section 'Participation'). In France, for example, despite discursive commitment to bottom-up governance, several studies (Ajates Gonzalez et al. 2018; Lamine 2017) note the dominance of large farmers' unions, public research, technical institutes and agricultural chambers in shaping and implementing EM policy. This is incompatible with political agroecology that aims to empower traditionally excluded and marginalized groups.

FEED THE WORLD

As a framing, the idea of feeding the world is often underpinned by an alarmist discourse on population growth, hunger and climate change. These serve to embed the emphasis on productivity as the key challenge in nourishing populations (IPES-Food 2016; Fouilleux et al. 2017).

The 'feed the world' frame is frequently used in conjunction with eye-catching statistics, also found in high-profile FAO publications, anticipating that world food production will have to increase by at least 50% by 2050 compared to 2012 levels, while in sub-Saharan Africa and South Asia, output will have to more than double (Tomlinson 2011). In China, Zhenzhong Si et al. (2018) argue that the 'feed China' narrative plays a similar role. This exclusive focus on short-term productivity almost entirely disables agroecology transformations by erasing multidimensional and long-term regenerative processes and functions. The 'feed the world' frame also promotes even more ecologically destructive production methods and downplays justice and distribution issues related to poverty and social inclusion, focusing even more intensely on industrialization and global trade as a means of addressing food insecurity.

Governments and private sector actors who deploy this frame often promote technological packages associated with the Green and Blue Revolutions, combined with liberalized international trade and underpinned by an ideological commitment to wealth and progress based on economic growth (IPES-Food 2016; Fouilleux et al. 2017; Tomlinson 2011).

Eve Fouilleux et al. (2017) note that social movements, particularly peasant groups, strongly oppose these mechanisms and 'solutions' but do not always disagree with the discourse on the need to produce more food. They frame family farming as the essential lever to nourishing local communities across the world, emphasizing not just productivity and availability but also food sovereignty, food and nutrition security, the right to food and food justice. However, the asymmetry in resources and access to powerful arenas where food policies are negotiated means that approaches aimed at increasing productivity prevail, thereby hampering agroecological transformations.

References

Ajates Gonzalez, R., Thomas, J., & Chang, M. (2018). Translating Agroecology into Policy: The Case of France and the United Kingdom. *Sustainability, 10*(8), 19.

Arnstein, S. R. (1969). A Ladder of Citizen Participation. *Journal of the American Institute of planners, 35*(4), 216–224.

Articulação Nacional de Agroecología (ANA). (2018). Sem Feminismo Não Há Agroecologia. *IV Encontro Nacional de Agroecologia (ENA), Belo Horizonte, Brasil.*

Baker, L., Gemmill-Herren, B., & Leippert, F. (2019). Accelerating Transformations to Sustainable Food Systems. In *Accelerating Transformations to Sustainable Food Systems: Beacons of Hope.* Biovision Foundation and Global Alliance for the Future of Food.

Benford, R. D., & Snow, D. A. (2000). Framing Processes and Social Movements: An Overview and Assessment. *Annual Review of Sociology, 26*(1), 611–639.

Bernard, B., & Lux, A. (2016). How to Feed the World Sustainably: An Overview of the Discourse on Agroecology and Sustainable Intensification. *Regional Environmental Change, 17*(5), 1279–1290.

Bezner Kerr, R., Hickey, C., Lupafya, E., & Dakishoni, L. (2019). Repairing Rifts or Reproducing Inequalities? Agroecology, Food Sovereignty, and Gender Justice in Malawi. *The Journal of Peasant Studies, 46*(7), 1499–1518.

Bhattacharya, N. (2017). Food Sovereignty and Agro-ecology in Karnataka: Interplay of Discourses, Identities, and Practices. *Development in Practice, 27*(4), 544–554.

Black Dirt Farm Collective. (2016). Afro-ecology: Cultivating Connectivity.

Chappell, M. J. (2018). Beginning to End Hunger: Food and the Environment in 332 Belo Horizonte, Brazil, and Beyond. Oakland: University of California Press.

Claeys, P. (2015). Food Sovereignty and the Recognition of New Rights for Peasants at the UN: A Critical Overview of La Via Campesina's Rights Claims over the Last 20 Years. *Globalizations, 12*(4), 452–465.

Cooke, B., & Kothari, U. (2001). *Participation: The New Tyranny?* London: Zed.

Desmarais, A. A. (2008). The Power of Peasants: Reflections on the Meanings of La Vía Campesina. *Journal of Rural Studies, 24*(2), 138–149.

Dryzek, J. S. (2013). *The Politics of the Earth: Environmental Discourses.* Oxford: Oxford University Press.

European Commission. (2018). *A Sustainable Bioeconomy for Europe: Strengthening the Connection between Economy, Society and the Environment.* Brussels: European Commission.

FAO. (2014). *Building a Common Vision for Sustainable Food and Agriculture. Principles and Approaches* (p. 50). Rome: FAO.

FAO. (2017). *Climate Smart Agriculture Sourcebook* (2nd ed.).

FAO. (2018a). *The 10 Elements of Agroecology.* Rome: FAO.

FAO. (2018b). *FAO's Work on Agroecology. A Pathway to Achieving the SDGs.* Rome: FAO.

FAO. (2018c). *Scaling up Agroecology Initiative: Transforming Food and Agricultural Systems in Support of the SDGs.* Rome: FAO.

FAO. (2019). Unlocking the Potential of Agricultural Innovation to Achieve the Sustainable Development Goals. In J. Ruane (Ed.), *Proceedings of the International Symposium on Agricultural Innovation for Family Farmers* (p. 120). Rome.

Fouilleux, E., Bricas, N., & Alpha, A. (2017). 'Feeding 9 billion people': Global Food Security Debates and the Productionist Trap. *Journal of European Public Policy, 24*(11), 1658–1677.

Friends of the Earth International. (2015). *Agroecology and Climate Justice: A People's Guide to Paris and Beyond.*

Geels, F. W., & Verhees, B. (2011). Cultural Legitimacy and Framing Struggles in Innovation Journeys: A Cultural-Performative Perspective and a Case Study of Dutch Nuclear Energy (1945–1986). *Technological Forecasting and Social Change, 78*(6), 910–930.

Giraldo, O. F., & Rosset, P. M. (2018). Agroecology as a Territory in Dispute: Between Institutionality and Social Movements. *The Journal of Peasant Studies, 45*(3), 545–564.

Global Forum for the Future of Agriculture (GFFA). (2020). Food for All! Trade for Secure, Diverse and Sustainable Nutrition. Communiqué 2020.

HLPE. (2019). *Agroecological and Other Innovative Approaches for Sustainable Agriculture and Food Systems that Enhance Food Security and Nutrition*. Rome: High Level Panel of Experts on Food Security and Nutrition of the Committee on World Food Security.

Holt-Giménez, E., & Altieri, M. A. (2013). Agroecology, Food Sovereignty, and the New Green Revolution. *Agroecology and Sustainable Food Systems, 37*(1), 90–102.

Horlings, L. G., & Marsden, T. K. (2011). Towards the Real Green Revolution? Exploring the Conceptual Dimensions of a New Ecological Modernisation of Agriculture that Could 'feed the world'. *Global Environmental Change, 21*(2), 441–452.

IPES-Food. (2016). From Uniformity to Diversity: A Paradigm Shift from Industrial Agriculture to Diversified Agroecological Systems. International Panel of Experts on Sustainable Food Systems (IPES).

IPES-Food. (2019). Towards a Common Food Policy for the EU.

Isgren, E. (2016). No Quick Fixes: Four Interacting Constraints to Advancing Agroecology in Uganda. *International Journal of Agricultural Sustainability, 14*(4), 428–447.

Khadse, A., Rosset, P. M., Morales, H., & Ferguson, B. G. (2017). Taking Agroecology to Scale: The Zero Budget Natural Farming Peasant Movement in Karnataka, India. *The Journal of Peasant Studies, 45*(1), 192–219.

Kothari, A., Demaria, F., & Acosta, A. (2015). Buen Vivir, Degrowth and Ecological Swaraj: Alternatives to Sustainable Development and the Green Economy. *Development, 57*(3–4), 362–375.

La Via Campesina. (2018). La Via Campesina in Action for Climate Justice. *Ecology, 44*(6), 1–32.

Lamine, C. (2017). *La fabrique sociale de l'écologisation de l'agriculture*. Marseille: Éditions la Discussion.

Lamine, C., Magda, D., & Amiot, M.-J. (2019). Crossing Sociological, Ecological, and Nutritional Perspectives on Agrifood Systems Transitions: Towards a Transdisciplinary Territorial Approach. *Sustainability, 11*(5), 1284.

Levidow, L. (2015). European Transitions Towards a Corporate-environmental Food Regime: Agroecological Incorporation or Contestation? *Journal of Rural Studies, 40*, 76–89.

Loconto, A. M., & Fouilleux, E. (2019). Defining Agroecology. *The International Journal of Sociology of Agriculture and Food, 25*(2), 116–137.

Maughan, C., Anderson, C., & Kneafsey, M. (2020). A Five-Point Framework for Reading for Social Justice: A Case Study of Food Policy Discourse in the Context of Brexit Britain. *Journal of Agriculture, Food Systems, and Community Development, 9*, 1–20.

Maughan, C., & Anderson, C. R. (Forthcoming). The Implications of the Innovation Agenda for Agroecology: A Case Study of the Agroecology Process at the U.N.

Méndez, V. E., Bacon, C. M., Cohen, R., & Gliessman, S. R. (2015). *Agroecology: A Transdisciplinary, Participatory and Action-oriented Approach*. Roca Baton: CRC Press.

Mier y Terán Giménez Cacho, M., Giraldo, O. F., Aldasoro, M., Morales, H., Ferguson, B. G., Rosset, P., et al. (2018). Bringing Agroecology to Scale: Key Drivers and Emblematic Cases. *Agroecology and Sustainable Food Systems, 42*(6), 637–665.

Nyeleni. (2015). Declaration of the International Forum for Agroecology.

Nyeleni Movement for Food Sovereignty. (2007). Nyéléni Declaration for Food Sovereignty.

Pimbert, M. P. (2015). Agroecology as an Alternative Vision to Conventional Development and Climate-Smart Agriculture. *Development, 58*(2), 286–298.

Pimbert, M. P. (2018a). Democratizing Knowledge and Ways of Knowing for Food Sovereignty, Agroecology and Biocultural Diversity. In M. P. Pimbert (Ed.), *Food Sovereignty, Agroecology and Biocultural Diversity. Constructing and Contesting Knowledge* (pp. 259–321). London: Routledge.

Pimbert, M. P. (2018b). *Food Sovereignty, Agroecology and Biocultural Diversity: Constructing and Contesting Knowledge*. London: Routledge.

Pimbert, M. P., & Borrini-Feyerabend, G. (2019). *Nourishing Life—Territories of Life and Food Sovereignty* (Policy Brief of the ICCA Consortium no. 6): The ICCA Consortium, Centre for Agroecology, Water and Resilience at Coventry University (UK) and CENESTA (Iran).

Rivera Ferre, M. G. (2018). The Resignification Process of Agroecology: Competing Narratives from Governments, Civil Society and Intergovernmental Organizations. *Agroecology and Sustainable Food Systems, 42*(6), 666–685.

Rosset, P. M., & Martinez-Torres, M. E. (2012). Rural Social Movements and Agroecology: Context, Theory, and Process. *Ecology and Society, 17*(3), 17.

Santos, B. d. S. (2015). *Epistemologies of the South: Justice Against Epistemicide*. New York: Routledge.

Schilling-Vacaflor, A. (2011). Bolivia's New Constitution: Towards Participatory Democracy and Political Pluralism? *Revista Europea de Estudios Latinoamericanos y del Caribe/European Review of Latin American and Caribbean Studies, 90*, 3–22.

Schmitt, C., Niederle, P., Ávila, M., Sabourin, E., Petersen, P., Silveira, L., et al. (2017). A experiência brasileira de construção de políticas públicas em favor da agroecologia. In *Políticas Públicas a favor de la Agroecología en América Latina y el Caribe*. Porto Alegre: Red Políticas Publicas en América Latina y el Caribe (PP-AL) & FAO.

Schneider, M. (2015). What, Then, Is a Chinese Peasant? Nongmin Discourses and Agroindustrialization in Contemporary China. *Agriculture and Human Values, 32*(2), 331–346.

Schwendler, S. F., & Thompson, L. A. (2017). An Education in Gender and Agroecology in Brazil's Landless Rural Workers' Movement. *Gender and Education, 29*(1), 100–114.

Shiming, L., & Gliessman, S. R. (2017). *Agroecology in China: Science, Practice, and Sustainable Management*. CRC Press.

Si, Z., Koberinski, J., & Scott, S. (2018). *Shifting from Industrial Agriculture to Diversified Agroecological Systems in China*. Report prepared for IPES Food.

Steinberg, M. W. (1998). Tilting the Frame: Considerations on Collective Action Framing from a Discursive Turn. *Theory and Society, 27*(6), 845–872.

Technical Centre for Agricultural and Rural Cooperation (CTA). (2019). The Digitalisation of African Agriculture Report 2018–2019. *Written by: Michael Tsan, Swetha Totapally, Michael Hailu and Benjamin Addom.* Netherlands: CTA.

Timmermann, C., & Félix, G. F. (2015). Agroecology as a Vehicle for Contributive Justice. *Agriculture and Human Values, 32*(3), 523–538.

Tomlinson, I. (2011). Doubling Food Production to Feed the 9 Billion: A Critical Perspective on a Key Discourse of Food Security in the UK. *Journal of Rural Studies, 29*(0), 81–90.

UNDROP, 2018. The United Nations declaration on the Rights of Peasants and Other People Working in Rural Areas [available at http://ap.ohchr.org/documenta/dpage_e.aspx?ai=A/hrC/39/L.16]

van der Ploeg, J. D., Barjolle, D., Bruil, J., Brunori, G., Costa Madureira, L. M., Dessein, J., et al. (2019). The Economic Potential of Agroecology: Empirical Evidence from Europe. *Journal of Rural Studies, 71*, 46–61.

van Walsum, E., Bruil, J., & Pascieznick, N. (2014). *Unlocking the Potential of Family Farmers with Agroecology* (pp. 42–45). Deep Roots, FAO and Tudor Rose.

Wezel, A., Bellon, S., Doré, T., Francis, C., Vallod, D., & David, C. (2009). Agroecology as a Science, a Movement and a Practice. A Review. *Agronomy for Sustainable Development, 29*(4), 503–515.

World Economic Forum (WEF). (2018). Innovation with a Purpose: The Role of Technology Innovation in Accelerating Food Systems Transformations.

World Forum of Fisher Peoples (WFFP). (2017). Agroecology and Food Sovereignty in Small-Scale Fisheries. Indonesia.

PART III

Drilling Down on Power and Governance in Agroecology Transformations

Within each of the domains discussed in Part II, we examined the dynamics, initiatives and approaches that enable agroecology, as well as the disabling factors that arise from the lock-ins, path-dependencies and power dynamics of the dominant regime. We now turn to a discussion of the relationship between the domains of transformation on the one hand and power and governance on the other, especially at the interface of the dominant regime and agroecology. We ask what type of governance works best to support system transformation, given the nature of agroecology.

To that end, we identify six different kinds of effects that governance and policy interventions can have on agroecology, from suppressing and undermining it to nurturing it and dismantling the regime, reflecting the systems transformation needed to support agroecology. These governance interventions take place at multiple scales, and may include policies but also starting a new network, developing markets, providing funding (philanthropic, private or public) to develop a new initiative, introducing new legislation, changing regulations, developing a specific learning process, starting a research project, developing a campaign for gender equity in agriculture—even persecuting activists. Each of these interventions, including those made by the state, private sector and civil society, is power-ridden, shaping the potential for agroecological transformations. From our analysis, it becomes clear how each of the six types of interventions can shape the enabling or disabling conditions for transformative agroecology within (and across) the domains.

The effects of these interventions are however not static or constant. For example, as we will discuss in the following chapters, interventions that were intended to support agroecology may end up co-opting or containing it. Processes of transformation are dynamic and the nuanced differences between progress and retrenchment are difficult to see in the moment or through a singular lens. This is why reflexive governance processes with the central involvement of food producers and other affected citizens are a central plank for agroecology transformations. This section presents a six-part framework to conceptualize the disabling and enabling effects of different governance interventions, what reflexive governance looks like and how it can be implemented at the pivotal territorial scale.

CHAPTER 10

Power, Governance and Agroecology Transformations

Abstract In this chapter, we focus on issues of power, control and governance in agroecology transformations. Synthesizing the findings across the six domains of transformation introduced in Part II, we explore how the different 'governance interventions' of different actors have multiple effects on a transformative agroecology. Interventions that undermine agroecology have two effects: (i) *suppressing* agroecology by actively repressing and criminalizing it and (ii) *co-opting* agroecology by supporting it only to become equivalent to the dominant regime. Interventions that *maintain the status quo* enable co-existence by (iii) *containing* agroecology as elements of the dominant regime are strengthened and alternatives ignored and (iv) *shielding* agroecology from regime dynamics so it is less threatened. In contrast, agroecological transformation of agri-food systems are enabled by (v) processes that *support* and *nurture* agroecology to develop on its own terms and (vi) *release* agroecology from its disabling context by dismantling elements of the dominant regime and *anchoring the values, norms and practices* of agroecology within and between territories, and at different scales.

Keywords Governance • Power • Intervention • Transformation • Social movements

© The Author(s) 2021
C. R. Anderson et al., *Agroecology Now!*,
https://doi.org/10.1007/978-3-030-61315-0_10

153

Fig. 10.1 Interventions can influence niche-regime dynamics in each of the domains in different ways. They can:

Strengthen the regime and undermine agroecology by:

Suppressing agroecology by actively repressing and criminalizing it

Co-opting agroecology by supporting it only to become equivalent to regime dynamics/values/norms

Maintain the status quo and enable co-existence by:

Containing agroecology by passively keeping it marginal as regime elements are strengthened and alternatives ignored

Shielding agroecology from regime dynamics so it is less threatened

Transform the regime and support agroecology by:

Nurturing agroecology to encourage its strengthening on its own terms

Release agroecology from its disabling context by dismantling

Starting from the basis of political agroecology, we looked at each domain to understand the dynamics of power and governance at play in the interface between agroecology and the dominant regime. The conditions in each domain provide an understanding of how change occurs—how enabling dynamics are bolstered and disabling ones are deconstructed and replaced. But we need to look at how we get from 'here to there'. Systems-wide transformation can only occur when the power dynamics that support the regime are confronted and when power shifts to excluded actors and groups. Peoples' agency is the crux of agroecology transformations, and it is on this that we now focus.

First, let us look at governance, which is often confused with government. Indeed, many proponents of agroecology work in institutions, governments and academia, a world where policy change is often viewed as the primary mechanism of social change. Yet our analysis suggests that it is important to move beyond the notion of policies as a starting point for the scaling of a transformative agroecology. A political agroecology is as much about the process, politics and principles of mobilization and shifting power.

The state—law, policy and regulations specifically—forms an often-problematic mechanism in the context of agroecological transformation (Giraldo and McCune 2019; Meek and Anderson 2020; see later section on oppressing and co-opting). Our view, experiences and review of the literature on agroecology transformations indicate that the locus of change from the incumbent regime to a transformative agroecology requires a much broader conceptualization of governance and policy. In any given place (a community, a city, a country, etc.), governance sets rules, rights of access, and the design of economic tools and accountability mechanisms for all actors involved. Governance determines how agroecology is supported as it spreads and how it is strengthened across sectors, regions and countries. Governance is about how power is exercised to take decisions (in communities, families, policy arenas, etc.) or not and who benefits and who does not from the process of change.

The forms of governance that prevail in each of the domains of transformation vary depending on the culture, history and balance of social forces in a particular context. In general, however, in the dominant regime, both the process and the resulting quality of governance will reflect and reinforce the interests of the powerful—be they political parties, elite groups, particular social formations (such as white men), influential families, large transnational corporations or global financial investors. For example, in communities where wages dictate access to food, location dictates access to education, and there are strong traditions of deference to certain types of people (e.g., gentry, men, elders, businessowners), these factors affect who is able to work on social change and participate in government (formal governance) or cultural education, protests and community organizing (informal governance).

Thus, policy is just one governance intervention among many. Other types of governance interventions can include building new market arrangements, learning strategies and decision-making protocols. As such, changes in policy, regulations and law are part of the tactical repertoire in the movement for agroecology transformations and must be approached cautiously and reflexively as a part of a wider set of actions for change. Many interventions, but especially policies, often fail to meaningfully shift power and become transformative—not just those by government but also those by civil society, including NGOs and social movements.

Elsewhere in this book, we discussed how agroecological transformations are non-linear, context-specific and messy processes that occur in the amorphous space between niche and regime. In this space, governance interventions play a crucial role. There is no single monolithic transformation unfolding in any one place. Indeed, the large-scale transformation of food systems is actually *many transformations*, in which cultural shifts, policy changes, struggles and networks intervene in complex, dynamic, often contradictory ways. It is vital to develop ways of thinking and working that help to confront, shape, improve and harness the "hopeful monstrosities" (Mokyr 1990, p. 291, cited in Schot and Geels 2008) that inevitably emerge in the imperfect process of transformation at different scales, places and times.

Our agency-centric approach calls attention to the capacity and necessity for hitherto excluded communities and social movements to build countervailing power and to collectively intervene in the governance of food systems. Fundamental to this process is the recognition that change requires both a diversity of tactics in different spaces and a capacity to be bold and reflexive at the same time. There are a range of ways of working for transformation—different strategies, tactics and theories of change—that cannot be universalized as appropriate or most effective in all possible contexts. What can be universalized is the elevation of social justice, sustainability and well-being as the central aims for governance interventions in agroecology transformations. Further, while we centre our analysis on the ability of collectives to intervene in governance and to exercise influence, we fully recognize that ultimately agroecological transformation cannot be controlled by individual actors or actions and thus transformations must be approached reflexively.

In summary, all these dimensions of power and governance permeate and shape the change that can take place in the domains and thus the quality and validity of agroecological transformations. It is vital to better understand the effects of interventions, in the governance realm of each domain of transformation as they shape the interaction between the regime and the agroecological niche—whether these interventions emanate from formal governments or from civil society.

Six Effects of Governance Interventions on Agroecological Transformations

How do different actors, collectives, institutions and governments intervene effectively across the six domains to support lasting and transformational agroecological change? In this section, we deepen the analysis of the

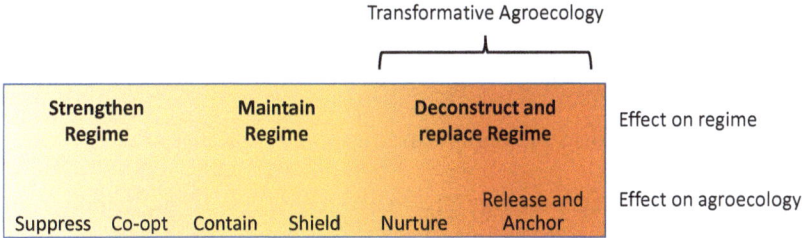

Fig. 10.2 Interventions can influence niche-regime dynamics in each of the domains in different ways

domains by discussing six different effects (Fig. 10.2) that interventions, including policies, can have in relation to agroecology transformations. Since transformation is a complex adaptive process, no one actor can determine it alone or *directly* by intentional interventions, including those involving financial or state resources. Interventions can, however, increase or decrease the likelihood of directions and outcomes, and sustained or repeated, concerted efforts at particular socio-spatio-temporal junctures make desired transformations much more likely. So, it is important to consider effects that interventions can have on agroecology—and how these may differ in terms of transformational capacity.

This section builds on and provides further nuance to earlier work on the binary between transforming and conforming to the dominant food regime (Levidow et al. 2014). We thus present three effects that undermine agroecology and reinforce the dominant regime (suppressing, co-opting and containing) and three that support agroecology transformations (shielding, nurturing and releasing/anchoring). Placing them along a spectrum, it becomes clear that two effects (containing and shielding) represent a middle ground that maintains the status quo and enables the dominant regime and agroecology to co-exist. As we will see, there is a fine line between the various effects, and they can change over time.

Interventions can strengthen the regime and undermine agroecology by:

- *Suppressing* agroecology by actively repressing and criminalizing it
- *Co-opting* agroecology by supporting it only to become equivalent to regime dynamics/values/norms

They can maintain the status quo and enable co-existence by:

- *Containing* agroecology by passively keeping it marginal as regime elements are strengthened and alternatives ignored
- *Shielding* agroecology from regime dynamics so it is less threatened

They can transform the regime and support agroecology by:

- *Nurturing* agroecology to encourage its strengthening on its own terms
- *Release* agroecology from its disabling context by dismantling elements of the regime and *anchor* agroecology by replacing these elements with niche dynamics, values, norms and practice

Interventions That Undermine Agroecology and Strengthen the Regime

Suppressing

The first of the three types of governance interventions that undermine agroecology involves *suppressing* people and processes that are favourable to agroecology. This can range from relatively non-violent actions to violent repression and persecution of advocates of agroecology.

Government funding programmes can intentionally bar small-scale producers, women, farmers from lower castes or other agroecological practitioners from government programmes. For example, in the UK Basic Payment Scheme, farmers with under five hectares of land have been ineligible for direct payment subsidies; yet smallholder farmers are key agents of agroecology in Britain and elsewhere. The state may also criminalize processes key to agroecology. For example, the implementation of seed laws that prevent replanting, and trading of farmers' seeds in communities and in peasant seed networks, is a significant impediment to agroecology, leaving food producers dependent on commercial seed suppliers (Goyes and South 2016). Similarly, food safety regulations that make traditional food processing illegal, often with no evidence of any added food safety risk, can stifle agroecology (Wallace 2016; Laforge et al. 2016). And there have been efforts to characterize local exchange and solidary economies, which are essential to agroecology, as a "barrier to trade" (see Chap. 6 on systems of economic exchange).

Interventions can also more actively dismiss agroecology in the discursive arena (see Chap. 9 on discourse). This has been seen more often in the past five years, in tandem with agroecology's rising profile at the Food and Agriculture Organization of the United Nations (FAO) and in the international community. For example, in a February 2020 speech and again in an August 2020 commentary, Kip Tom, US permanent representative to FAO, claimed that agroecology was responsible for recent locust infestations in Africa. He attacked agroecology, through recurring oversimplifications of the approach as rejecting "20th century technologies that undergird food security", being 'anti-mechanization' and even intentionally being used by European interests to prevent "progress" in Africa (Tom 2020). Part of Tom's role is advancing US business interests, which are undermined by political agroecology. As he noted towards the end of his speech, his agenda is: "We want to make sure we're delivering value that allows for the free trade of America's products around the world." Similar attacks on agroecology have been made by other industry apologists who have also written op-eds in the farm and popular media.

Suppression can also be physically and brutally violent. Historically, such violence was widely accepted as a basis for colonial expansion—for example, through the dispossession of indigenous and traditional peoples (often historically practising vernacular forms of agroecology) and the installation of colonial monoculture plantation structures and export-focused, capital-intensive regimes. Today, this violent neo-colonial dynamic is more implicit and is enacted through state-, corporate- and big NGO-sponsored projects and programmes.

Increasingly, activists and land defenders who interfere with state-led and corporate-backed development agendas find themselves victims of such violence. In his 1997 book *Green Backlash*, Andrew Rowell provided one of the first significant accounts of the tactics through which governments, corporations and authoritarian factions of society actively subverted the environmental movement, in part through direct physical violence against environmental activists, indigenous peoples and other groups. A prominent example is the forced eviction of people from their land—through land grabs—for private, state or conservation programmes. A more subtle form of violence may take place, when governments and local elites urge resettlement as the only alternative (Milgroom and Spierenburg 2008).

Other interventions that suppress agroecology are the intimidation and murder of ordinary people challenging the dominant regime, demanding social justice or defending the environment, the peasantry or rural liveli-hoods. For example, Global Witness (2018) documented the murder of 207 people in 2017 who stood up to governments and companies—and these are only the documented cases. Their report describes how agribusi-ness "was the most dangerous sector, overtaking mining for the first time ever, with 46 defenders killed protesting against the way goods we con-sume are being produced" (Global Witness 2018, p. 8). In some cases, this violence directly involves state actors—police, military or paramilitary forces. In others, private security forces of corporations have carried out the killings while the state acts with indifference towards such violence.

Today, with many national governments exhibiting increasingly author-itarian tendencies, the suppressive effect of the dominant regime is likely to grow. Omar Felipe Giraldo and Nils McCune (2019) detail this in a recent article, noting that "not only have agroecology-friendly policies been overturned in these countries, but they have arguably been con-verted into tools for repression and information-gathering against move-ments". One example of this is from the Brazilian programme for the public procurement of food from family farmers (PAA). In the hands of the Temer government (2016–2018), data on cooperatives became a tool for judicial harassment against local farmer organizations, while from 2019 onwards the Bolsonaro administration has been dismantling policies that support agroecology (Borborema 2020) and even shut down the entire Ministry for Agrarian Development—which until then was a key institutional basis for the promotion of agroecology in the country.

Finally, this violence is inflicted not only on people but also on the eco-logical, cultural and material basis of their identity and livelihoods, further suppressing agroecology. For example, in the mid-2000s in Guatemala, "a boom in environmentally harmful extractive industries brought about by deregulation, World Bank loans to build extractive infrastructure, and increased global demand for land-based resources" (Copeland 2018) led to substantial ecological deterioration, deeply affecting local and indige-nous communities. Governance interventions that enact ecological vio-lence, reduce access and undermine rights to nature (including land, seeds, water and ecosystems, as in Domain A (Chap. 4))—or that facilitate their violent enclosure by large corporate interests—are some of the most debil-itating factors for agroecology and agroecology transformations.

Co-opting

Governance interventions in any of the domains of transformation can also have the effect of incorporating or *co-opting* agroecology in ways that encourage conformity to the dominant regime. Rob Raven and Adrian Smith (2012) refer to a "fit and conform" mode of development where radical alternatives in niches, such as agroecology, are subjected to pressure (through incentives) to align with the tenets of the regime. The regime's underlying social, political and ecological processes meanwhile are not substantially affected (also see Levidow et al. 2014). When the actors, policies and mechanisms of the dominant regime take up agroecology it may lead to support, uptake and growth. But there is also the risk that agroecological systems and practices are recast in the mould of the regime and that the social and political forces of agroecology are neutralized and demobilized (van der Ploeg 2018, #4269; Giraldo and Rosset 2018).

In this respect, many government programmes and other interventions can intentionally or inadvertently appropriate practices, markets and discourse of agroecology, which in their original form were alive with transformative potential. A dynamic of co-optation emerges when interventions to foster agroecology only promote certain incremental changes—those that do not alter power relationships, that support technical fixes in reductionist ways, and that otherwise lack a holistic approach (across all domains) to enable social, cultural, political, economic and ecological dimensions of transitions to sustainable food systems. In other words, agroecology is co-opted where funding, support and extension programmes for agroecology bolster actors who already hold power in the dominant system—for example, where public resources for agroecology are allocated to corporate retailers or large farmers (Laforge et al. 2016). Again, this signals the importance assessing of who has power in processes of governance.

In some cases of co-optation, agroecology has been integrated into an actor's language, programmes and policies (e.g. in the case of a state or a development institution), but actual implementation is minimal or only marginally reflects true agroecology transformation. In some parts of Latin America, such as Ecuador, the promotion the principles of agroecology was enshrined in the constitution, yet substantial change in regime configurations—such as a meaningful shift from state support for export-oriented agriculture to peasant farming or agroecology—did not transpire (e.g. Intriago et al. 2017). Another example is the case of France, where agroecology has been adopted as a central plank of the Ministry of Agriculture's remit. This has been celebrated but also viewed with scepticism by many of the country's small-scale agroecological producers and others, who have observed that many of the government's programmes reflect a view of agroecology that only involves minor adjustments to an otherwise industrialized system.

Indeed, as Paulo Petersen et al. (2013) warn, the institutionalization of agroecology in government policy often leads to co-optation. It tends to reproduce the top-down approaches characteristic of industrial agricultural development, negating the agency of producers and producers' organizations. According to Stephen Sherwood et al. (2018), this reflects a "scaling up in name but not in meaning": many of the core principles of agroecology are sidelined in favour of ones that are more compatible with the status quo.

This process of co-optation is not only something people and institutions of the regime do to agroecology. It can be a more dispersed and subtle process, often perpetuated by well-meaning researchers governments or NGOs that promote agroecology but seek to mainstream it as a way of scaling it up. They may encourage a technocratic and compromised agroecology to gain institutional uptake (Box 10.1). This dynamic is why social movements and other protagonists of agroecology are putting great efforts into defining a transformative agroecology (Giraldo and Rosset 2018; Anderson et al. 2015) while rejecting diluted interpretations of agroecology that risk undermining their efforts.

The concept of agroecology can also be co-opted within the frame of commercialization and entrepreneurialism in agriculture, where it can be interpreted in conventional, 'regime oriented', ways (Isgren and Ness 2017) or become part of a 'corporate-environmental food regime' (Levidow 2015). For example, agroecological methods, but not necessarily principles, have been adopted by agrochemical companies, some governments and large-scale industrial producers who have incorporated agroecological techniques into 'sustainable intensification' agendas. In Europe, for instance, this nascent neo-productivist agenda selectively incorporates agroecological practices within a toolkit that also includes and promotes biotechnology (Levidow 2015). Such a move and process has been criticized by many farmers' organizations, NGOs and social movements, again reflecting the importance of social movements in contesting co-optation.

Box 10.1   'Best Policy for Agroecology' in Sikkim—a Deeper Look
Sikkim won many accolades for becoming the first Indian state to certify all agricultural production as organic by international standards in 2016, including the 2018 Future Policy Award for Best Policy Promoting Agroecology. However, a recent paper shows (Meek and Anderson 2020) that while Sikkim's policies may be celebrated as nurturing agroecology, the effects of its governance interventions contradict many of the key ecological, social and political principles of agroecology—especially regarding biodiversity, inclusion, farmer-agency and food sovereignty. "While in some cases the Sikkimese state may be encouraging integrated farming systems, their main thrust is advancing monocultural organic production for export production in a market [led] rather than [a] livelihood led approach" (Meek and Anderson 2020).

The idea of 'Organic Sikkim' is, in principle, congruent with agroecology. But it is a top-down approach that hardly takes account of the traditions, knowledges, genetic resources or the roles of local farmers and citizens, including in the formulation of policies. Yet Sikkim is home to rich agroecological traditions, "knowledges, wisdoms and community dynamics" that have been "an important contributor to food security, biodiversity and well-being and an ongoing field of potential in Sikkim to nurture a just, sustainable and culturally appropriate scaling process" (p. 17).

The government of Sikkim sees small land holdings in the mountains as limiting productivity and market potential. In order to address these perceived constraints, it is imposing new forms of social organization: constructing cooperatives and value chains from the top down. However, these state-led processes tend to benefit those who already hold power and to homogenize the hitherto diverse agricultural landscape to produce volumes of marketable goods and in the end consolidate power and capital and reduce biodiversity. Further, farmers have very little say in these policies, and the political dimensions of agroecology—especially the notion of centring the agency of food producers—are almost completely absent.

This case also highlights how any 'successful' example of agroecology should always be analysed with nuance and depth and it is necessary to evaluate critically to what extent governance interventions actually nurture agroecology. Such an approach helps to demystify the contextualized successes of interventions like Organic Sikkim and what they mean for other places looking to adopt policies for agroecology.
Source: Anderson and Meek (2020)

INTERVENTIONS THAT MAINTAIN THE STATUS QUO
AND ENABLE CO-EXISTENCE

Containing

The third disabling effect that governance interventions may have is containing agroecology, or passively keeping efforts to establish it from developing further. While there are a range of 'lock-in' mechanisms that keep the industrial food system in place, these simultaneously 'lock-out' or contain agroecology.

For example, in Chap. 6 (on the systems of economic exchange domain), we discussed how the requirement for standardized, high-volume products in conventional retail chains locks agroecological farmers out of these markets. In Chap. 5 (on the knowledge and culture domain), we showed that the non-recognition of local and traditional knowledge in mainstream science and the narrow economic and productivist basis of indicators marginalize agroecology. Another example is the failure of governments to curtail the concentration and power of corporate retailers, processors and suppliers of external inputs, which drastically contains the potential for agroecology to develop.

Interventions also contain agroecology by enabling actors within the dominant regime to gain yet more access to resources such as funding programmes for large-scale producers. In other instances, containment may arise from food safety regulations designed for large-scale processing and export markets. These are not practically or economically accessible for small agroecological producers or processors; the burden of compliance is only viable when spread across a large-scale operation. Policies or laws, and their implementation, may be inherently biased against women or other non-dominant groups. Those who do not actively take an equity-sensitive approach are highly likely to reinforce inequality and power imbalances, containing the emancipatory dimension of agroecology.

Similarly, discourses that position corporations and technology as the protagonists of sustainability transitions—such as in the climate-smart agriculture approach—are often amplified by powerful philanthropists such as the Bill and Melinda Gates Foundation and undermine an agroecology approach that centres the agency of food producers and nature. Support for industrial agriculture by governments, for example through providing subsidies for synthetic fertilizers, can lock farmers into such practices and out of agroecology. All these examples show how governance interventions can contain the potential for agroecology to contribute to the transformation towards sustainable food systems while keeping the dominant regime in place (Pimbert 2015).

Shielding

The first enabling governance intervention is *shielding*, where policy can help to protect agroecology from the most damaging pressures of the regime without changing the regime itself or directly providing resources to agroecology. As such, it is unlikely to significantly advance agroecology transformations in itself and largely maintains the status quo in terms of transformative potential. Shielding is often introduced through legislative and regulatory exceptions.

In environments where price and convenience are the main criteria for selecting suppliers, small-scale producers are drastically disadvantaged, given the economies of scale and standardization that large corporate food providers can attain. Governments can choose to provide relief, for example in food safety compliance, by not subjecting small-scale agricultural producers to the same requirements as large processors. For example, in British Columbia, Canada, Christiana Miewald et al. (2013) illustrated how important exemptions from industrial food safety standards are for small-scale meat processing: when the exemption was removed, the industry collapsed. Small-scale food producers can also be shielded from imbalanced trade environments through preferential access to institutional food buyers, as was the case with Brazil's School Feeding programme.

Although not specifically targeted at agroecology, green belts can help shield agroecology in peri-urban areas from displacement by development. Legislation that protects peasant and community seed networks and open source intellectual property can protect these important aspects of agroecology from the commercial interests of seed companies (García López et al. 2019). Trade rules that prevent the dumping of cheap imported food, which distorts markets and undermines agroecology, are also an example of shielding.

In essence, shielding is important because it can provide safe spaces for agroecological experiments to mature and protect them from being directly undermined. However, this shielding does little *per se* to actively enable agroecological transformation.

Interventions That Support Transformative Agroecology

Nurturing

Among transformative governance interventions, the active nurturing of political agroecology—through resources to facilitate the development of agroecological initiatives, networks, markets, innovations and more—is centrally important. Nurturing of agroecology involves support that adheres to the principles of a political agroecology, especially the idea of bolstering the agency of food producers, democratic governance and food sovereignty. In nurturing interventions, the allocation and use of any resources, internally or externally derived, are self-determined by agroecological producers, communities and territorial networks rather than by elite policy-makers or donors.

Nurturing by governments, social movements researchers or other actors involves providing funding, networks or technical support for agroecology initiatives and experiments. These may include cooperative processing schemes; bottom-up and multi-actor networks, platforms and organizations; participatory agroecological research and knowledge exchange; publicity; the development of nested markets; and initiatives that empower women and youth in agriculture. The support is explicitly aimed at enabling agroecology on the terms of its main protagonists. Rarely, but in an ideal case, support structures and priorities from governments and donors are developed through participatory decision-making and guided through the democratic participation of food producers.

In most cases, nurturing is not driven primarily by 'financing' from mainstream institutions and donors but rather is resourced by the social movements, communities and networks that collectively mobilize knowledge, resources and energies at different scales. The global network of agroecology schools supported by the transnational social movement La Via Campesina (LVC) is a wonderful example of one approach to nurturing. While the resources of the state have been leveraged to some degree, both LVC and the agroecology schools have self-organized in relative autonomy and sometimes opposition to the state.

In the United Kingdom, the work of the Landworkers' Alliance is exemplary, where a young and scrappy union of farmers, growers, foresters and land-based workers grew from a handful of members in the early 2010s to a powerful organization providing technical and political trainings on agroecology, mutual support networks, mentoring, web-based forums, farmer-led research and more. They also have provided leadership on campaigns to support networks that tackle multiple dimensions of inequity in the food system including gender, racism, immigration and they focus on supporting young farmers. This has also been accomplished through their embeddedness in European and global networks of organizations working to advance food sovereignty and agroecology. This has allowed for the cross-pollination of ideas from different places and also across boundaries, for example in their partnership with allied academic institutions to support their training, research and innovation agenda. The importance of local organizations in nurturing agroecology cannot be understated and further examples can be seen throughout Part II across all regions of the world.

Government involvement in nurturing agroecology generally involves a *co-existence* approach, where agroecology is intentionally supported by the state through shielding and nurturing but its further development is contained because of the simultaneous, dominant promotion of large-scale, external input-intensive agricultural production. Brazil, India, Nicaragua, Ecuador and a number of other countries are highlighted in the literature as examples where agroecology was supported by state policy, alongside a 'bigger sister' of export-led agriculture. In such cases, even with modest support from the state, small-scale farmers compete on unequal footing because disabling dynamics described in each of the domains are raising barriers (Mier y Terán Giménez Cacho et al. 2018).

The approach of co-existence could be viewed in two ways. On the one hand, it can be seen as creating further space in which agroecology can be strengthened; this could give it more potential to influence and transform the regime. At the same time, this two-tiered approach may represent a way to maintain agroecology as a permanent system next to other dominant systems that emphasize specialization, productivity, scale intensification and narrow technocratic solutions. But in the latter case, agroecology becomes confined or 'niche-ified' as a small sub-sector of agriculture rather than a viable vision for agricultural development.

Civil society is more often the driver in nurturing agroecology, because governmental involvement in it is so often fraught—the result of a sustained struggle and emerging political opportunities, where social movements have been able to anchor agroecology in the dominant regime (see later section on anchoring). Unfortunately, government support and short-term political cycles are not conducive to supporting the long-term transformation required for agroecology. Paulo Petersen et al. (2013, p. 111) argue that the "combination of the fragmentation of policies in space (the focus on administrative sectors) and time (the focus on the short term) imposes serious obstacles to the transition of public institutions from the perspective of agroecological development".

There is thus a need to look beyond election cycles and sectoral boundaries to consider a process of trans-local, continuous agroecological development. Nurturing agroecology transformations using the territorial approach (see Section 1: The Territorial Governanceof Agroecology Transformations in Chap. 11) can help to break with the sectoral reductionism that predominates in public policies and many aspects of food systems. Even in the most exemplary cases of governmental support for agroecology, the effects can often be fleeting. Changes in political winds can easily wipe out progress—as we are now witnessing in Brazil, where decades of nurturing agroecology, led by social movements and supported by governments, are being undone.

There is a fine line between nurturing and co-opting. Governance interventions by mainstream actors may appear supportive of a political agroecology, but often they are closely tied to wider economic development trajectories reflecting values of the dominant regime. In this case, programmes for agroecology that appear at first to be enabling may in effect encourage narrowing and specialization (rather than diversification), marketization (orienting towards a market only rather than the multiple functions of agroecology) and technocratic solutions (obscuring the political); they may ultimately leave out the emancipatory dimensions of agroecology (Laforge et al. 2016). Such dynamics can co-opt agroecology. Thus, while the pro-active nurturing of bottom-up forms of agroecology is vital for transformation, equally critical are interventions that directly contest and infiltrate the dominant regime.

Release and Anchor

Interventions can also have the effect of *dismantling the regime* to release agroecology from the disabling conditions it creates, while simultaneously *anchoring agroecology to the regime*, thereby transforming it. To make this possible, overt political contestation of the dominant regime is fundamental. Whereas nurturing is a constructive mode of working, releasing agroecology is deeply deconstructive for the regime. Transformations towards a political agroecology will be impossible if the dominant regime is not extensively gutted, for example by diverting state resources away from industrial agriculture towards agroecology or by replacing discourse that favours specialization and exports to one that emphasizes the multifunctionality of agroecology and local and territorial markets.

Efforts to release agroecology address how and why agroecology is contained, co-opted and suppressed—effects 1, 2 and 3 (above)—and instead deconstruct aspects of the regime to make it less hostile to agroecology. It is reflected in the work of social movements, activists, land defenders, critical scholars and investigative journalists that are advancing the material and discursive work of critique, dissent and protest. It is also reflected in campaigns against land grabbing and genetic modification that call out the co-optation of agroecology and fight for the break-up of corporate power in the food chain.

A few examples demonstrate some of these efforts in more detail. In the first example, a range of organizations is actively contesting the corporate-led roll-out of a new green revolution for Africa which promotes biotechnology and agri-industry as a solution to food in security in Africa. Recent reports (e.g. Wise 2020) have exposed the failures of these programmes to deliver on the food security outcomes that organizations like the Alliance for a Green Revolution in Africa (AGRA) claim to produce. They also demonstrate the ecological, economic and political downside to the AGRA approach for local peoples, including highlighting the many shortcoming of the first green revolution (in South Asia) as well as unveiling the vested interests of the corporate actors in the global north that finance and champion the green revolution approach.

Elsewhere, successful efforts to ban genetically modified organisms, or glyphosate in municipalities or at a national level, have resulted from the work of social movements in courts, legislatures and on the streets (Peschard and Randeria 2020). These wins are a part of the process of deconstructing the regulatory and legal frameworks that have enabled the encroachment of monocultural chemical agriculture into these territories, thus opening new possibilities for the emergence of alternatives. In another example from Manitoba, Canada, activists mobilized to contest the role of food safety regulations in containing the development of local sustainable food systems, arguing for scale-appropriate regulations and for policy that legitimately supported small farmers to develop agroecological food systems (Laforge et al. 2016).

The other dynamic, *anchoring*, occurs when niche-regime interactions lead to a durable connection between the two, creating new possibilities for change (Elzen et al. 2012). Anchoring of the niche to the regime can take place by establishing new rules or institutions, fostering new practices, processes or technologies, or building new networks and social groups where the tenets of agroecology legitimately infiltrate the regime (Schiller et al. 2019). Anchoring involves the engagement of agroecology proponents, organizations and social movements with the technological, network-related and institutional aspects of the regime. Often, entry points or bridges are found, which allow previously excluded agroecological actors and perspectives to gain access to institutional processes and networks where rules, legislation, discourse, norms and access to finance are negotiated.

Some examples of anchoring can be found at all levels of governance. At the international level, actors have been engaged in a longstanding struggle to anchor agroecology in FAO as an alternative to the dominant focus of a green revolution approach. After a decade of advocacy, negotiations and social movement struggle, FAO launched a global dialogue on agroecology between 2015 and 2018, culminating in the announcement of the Scaling Up Agroecology initiative and the commitment of FAO resources to support agroecology. In this case, a transformative and political agroecology became anchored in one of the most important global public institutions for shaping agricultural research, policy and discourse (Loconto and Fouilleux 2019). Similar work at the global level by social movements can be observed in the efforts to ratify new human rights, for example in the recently adopted declaration of peasant rights or in the Voluntary Guidelines on the Responsible Governance of Tenure of Land, Fisheries and Forests. These global-level anchoring processes create frameworks, mechanisms and resources that actors at local, territorial and national levels can draw on to advance agroecology—for example by holding states accountable to uphold peasants' rights or to implement responsible governance of tenure.

Anchoring can also be viewed in the form of local food policy councils or biodistricts (see Box 11.2 in Chap. 11) which, while varied in their form, can help to foster new markets, regulatory change and other dynamics that anchor agroecology. In their most transformative form, these include participatory dynamics that enable citizens to gain agency in policy-making.

In another example, the research centre where many of the authors of this book are embedded represents a form of anchoring. In 2015, the Centre for Agroecology, Water and Resilience (CAWR) was launched based on a transformative vision of agroecology that brings people's knowledge into dialogue with scientific knowledge in a transdisciplinary approach. CAWR is the largest centre of its kind focusing on agroecology and involves approximately 100 members including research staff, students and a support team. The centre has gone on to engage in and advocate for farmer-led, transdisciplinary and participatory approaches to research and food system governance and to focus on enabling agroecology in different territories.

Whereas many contributors to the multi-level perspective literature are rather sanguine about the potential of these anchoring processes, proponents of a political agroecology have been much more cautious and critical as they enter into these institutional spaces. Indeed, while these certainly can contribute to a transformative agroecology, anchoring processes can—like nurturing—also lead to compromises and concessions and ultimately morph into co-optation. This dynamic can be viewed in all of the examples provided in the preceding paragraphs, where the gains made by anchoring agroecology have been accompanied by dynamics (in FAO, in local policy councils and in research centres like CAWR) that reflect aspects of the dominant regime and that do little to substantially shift power. Transformations, again, are not linear and the dominant actors in intergovernmental institutions, local councils and academia—like in all spaces—adapt to claim back power and to supress and contain efforts to anchor agroecology.

All of these examples highlight the dialectic and non-linear process of transformation and the challenge of anchoring agroecology in the regime. In the next chapter we will discuss how continuous reflection on the effects of any intervention is key to ensure a transformative agroecology and how this is best done by the key actors of a given agroecological territory in participatory processes.

References

Anderson, C. R., Pimbert, M., & Kiss, C. (2015). Agroecology as Practice, Movement and Vision. https://www.youtube.com/watch?v=-Km9Kv5UylU&feature=youtu.be

Borborema, H. (2020). Bolsonaro veta quase todos os artigos do projeto de lei de apoio à agricultura familiar. Available: https://agroecologia.org.br/2020/08/25/bolsonaro-veta-projeto-de-lei-de-apoio-a-agricultura-familiar/.

Copeland, N. (2018). Meeting Peasants Where They Are: Cultivating Agroecological Alternatives in Neoliberal Guatemala. *The Journal of Peasant Studies, 46*(4), 831–852.

Elzen, B., van Mierlo, B., & Leeuwis, C. (2012). Anchoring of Innovations: Assessing Dutch Efforts to Harvest Energy from Glasshouses. *Environmental Innovation and Societal Transitions, 5*, 1–18.

García López, V., Giraldo, O. F., Morales, H., Rosset, P. M., & Duarte, J. M. (2019). Seed Sovereignty and Agroecological Scaling: Two Cases of Seed Recovery, Conservation, and Defense in Colombia. *Agroecology and Sustainable Food Systems, 43*(7–8), 827–847.

Giraldo, O. F., & McCune, N. (2019). Can the State Take Agroecology to Scale? Public Policy Experiences in Agroecological Territorialization from Latin America. *Agroecology and Sustainable Food Systems, 43*(7–8), 785–809.

Giraldo, O. F., & Rosset, P. M. (2018). Agroecology as a Territory in Dispute: Between Institutionality and Social Movements. *The Journal of Peasant Studies, 45*(3), 545–564.

Global Witness. (2018). At What Cost? Irresponsible Business and the Murder of Land and Environmental Defenders in 2017.

Goyes, D. R., & South, N. (2016). Land-grabs, Biopiracy and the Inversion of Justice in Colombia. *The British Journal of Criminology, 56*(3), 558–577.

Intriago, R., Gortaire Amézcua, R., Bravo, E., & O'Connell, C. (2017). Agroecology in Ecuador: Historical Processes, Achievements, and Challenges. *Agroecology and Sustainable Food Systems, 41*(3–4), 311–328.

Isgren, E., & Ness, B. (2017). Agroecology to Promote Just Sustainability Transitions: Analysis of a Civil Society Network in the Rwenzori Region, Western Uganda. *Sustainability, 9*(8), 1357.

Laforge, J. M. L., Anderson, C. R., & McLachlan, S. M. (2016). Governments, Grassroots, and the Struggle for Local Food Systems: Containing, Coopting, Contesting and Collaborating. *Agriculture and Human Values, 34*(3), 663–681.

Levidow, L. (2015). European Transitions Towards a Corporate-environmental Food Regime: Agroecological Incorporation or Contestation? *Journal of Rural Studies, 40*, 76–89.

Levidow, L., Pimbert, M., & Vanloqueren, G. (2014). Agroecological Research: Conforming—or Transforming the Dominant Agro-Food Regime? *Agroecology and Sustainable Food Systems, 38*(10), 1127–1155.

Loconto, A. M., & Fouilleux, E. (2019). Defining Agroecology. *The International Journal of Sociology of Agriculture and Food, 25*(2), 116–137.

Meek, D., & Anderson, C. R. (2020). Scale and the politics of the Organic Transition in Sikkim, India. *Agroecology and Sustainable Food Systems*, 1–20. https://doi.org/10.1080/21683565.2019.1701171.

Mier y Terán Giménez Cacho, M., Giraldo, O. F., Aldasoro, M., Morales, H., Ferguson, B. G., Rosset, P., et al. (2018). Bringing Agroecology to Scale: Key Drivers and Emblematic Cases. *Agroecology and Sustainable Food Systems, 42*(6), 637–665.

Miewald, C., Ostry, A., & Hodgson, S. (2013). Food Safety at the Small Scale: The Case of Meat Inspection Regulations in British Columbia's Rural and Remote Communities. *Journal of Rural Studies, 32*, 93–102.

Milgroom, J., & Spierenburg, M. (2008). Induced Volition: Resettlement from the Limpopo National Park, Mozambique. *Journal of Contemporary African Studies, 26*(4), 435–448.

Peschard, K., & Randeria, S. (2020). 'Keeping seeds in our hands': The Rise of Seed Activism. *The Journal of Peasant Studies, 47*(4), 613–647.

Petersen, P., Mussoi, E. M., & Dal Soglio, F. (2013). Institutionalization of the Agroecological Approach in Brazil: Advances and Challenges. *Agroecology and Sustainable Food Systems, 37*(1), 103–114.

Pimbert, M. P. (2015). Agroecology as an Alternative Vision to Conventional Development and Climate-Smart Agriculture. *Development, 58*(2), 286–298.

Schiller, K., Godek, W., Klerkx, L., & Poortvliet, P. M. (2019). Nicaragua's Agroecological Transition: Transformation or Reconfiguration of the Agri-food Regime? *Agroecology and Sustainable Food Systems, 44*, 1–18.

Schot, J., & Geels, F. W. (2008). Strategic Niche Management and Sustainable Innovation Journeys: Theory, Findings, Research Agenda, and Policy. *Technology Analysis & Strategic Management, 20*(5), 537–554.

Sherwood, S., Arce, A., & Paredes, M. (2018). Affective Labor's 'unruly edge': The Pagus of Carcelen's Solidarity & Agroecology Fair in Ecuador. *Journal of Rural Studies, 61*, 302–313.

Smith, A., & Raven, R. (2012). What Is Protective Space? Reconsidering Niches in Transitions to Sustainability. *Research Policy, 41*(6), 1025–1036.

Tom, K. (2020). Speech at U.S. Department of Agriculture's (USDA) 2020 Agricultural Outlook Forum.

van der Ploeg, J. D. (2018). *The New Peasantries: Rural Development in Times of Globalization.* Earthscan Food and Agriculture.

Wallace, R. (2016). *Big Farms Make Big Flu: Dispatches on Influenza, Agribusiness, and the Nature of Science.* New York, NY: Monthly Review Press.

Wise, T. (2020). False Promises: The Alliance for a Green Revolution in Africa (AGRA). https://www.rosalux.de/fileadmin/rls_uploads/pdfs/Studien/False_Promises_AGRA_en.pdf

CHAPTER 11

Reflexive Participatory Governance for Agroecological Transformations

Abstract In this chapter we further discuss the rationale for a participatory and reflexive governance process as the basis for agroecology transformations. We discuss governance and facilitation mechanisms that enable continuous discussions, negotiations, exchange and joint planning between actors. Further, we provide guidance on this ongoing and iterative social learning processes among actors that can enable and ensure governance interventions that both nurture and anchor agroecology. This often requires an expansion of 'direct' democracy in decision-making in order to complement, or replace, models of representative democracy that prevail in conventional policy-making. Finally, we articulate the territorial approach to governance which is increasingly seen as the decisive level in fostering agroecological transformations and the scale where reflexive and participatory governance can be effectively implemented.

Keywords Participatory processes • Deepening democracy • Governance • Food sovereignty

As we have seen in the analysis of the domains, agroecological transformation emerges through collective action, often driven by the collective agency of food producers in territories. Its governance must be participatory rather than top down, for two reasons.

© The Author(s) 2021
C. R. Anderson et al., *Agroecology Now!*,
https://doi.org/10.1007/978-3-030-61315-0_11

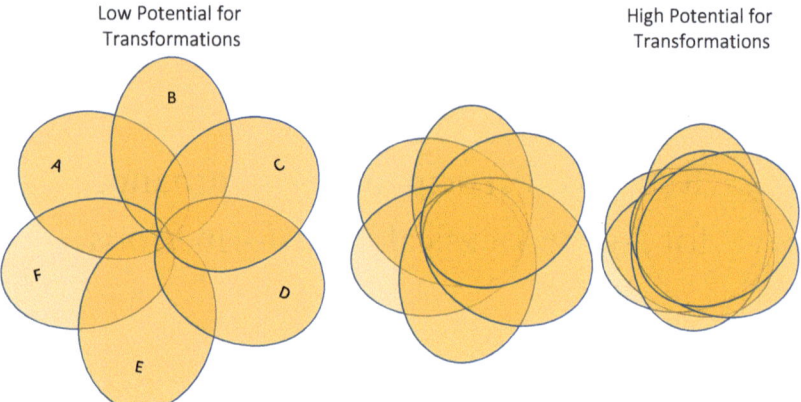

Fig. 11.1 On the left side, domains largely reflect disabling conditions for agro-ecology. As domains start to overlap, enabling conditions in each domain become more aligned, enhancing the potential agroecological transformation (right)

First, agroecology is generally based on universal *principles* or *elements* that demand adaptation to local context, not adoption of prescribed technological packages. In specific territorial contexts, these principles must be implemented in a way that reflects the social, political and biocultural contingencies and knowledge of place. In the same way, the domains we have presented offer not a prescription of interventions or technologies but rather suggestions of critical areas for the development of agroecology. Analysis and interventions must be articulated and deployed democratically in place for each domain.

It is very important to also consider the intersections and overlaps between domains. Generally, when governance interventions shift power in more than one domain (through one of the six effects described in earlier chapters), the possibilities for transformation increase. When processes of transformation within multiple domains start to overlap and become 'tied', the opportunities for wider transformation—in a locale, territory or a country—are amplified, as change in the domains tend to align and mutually reinforce each other. An integrative approach that addresses and ties these domains is impossible through the interventions of individual groups, government agencies and other actors operating in isolation. It is through participatory and democratic collective processes of negotiation, reflexive analysis and action in territories that this becomes possible. Figure 11.1 visualizes this confluence.

The second point relates to the effects of interventions on transformation processes. As we have seen, the direct effect(s) of interventions in the realm of governance cannot always be predicted. Most interventions are implemented in highly complex, relational situations that are in constant flux, and most often, it is only apparent in retrospect what influence a particular intervention has on the development of agroecology. As we have seen in the previous chapter, an intervention that had the intention of nurturing agroecology could evolve into dynamics that enable co-optation or containment instead—and vice versa, as, for example, in the case of Nicaragua (Box 11.1).

Box 11.1   Peace in Our Places of Origin: From Oppressing to Anchoring Agroecology in Nicaragua

Nicaragua's history has been a crucial factor in its agroecological transformation. The institutional environment created through social movement mobilization of the 1990s, and then transformed over three consecutive Sandinista governments, has favoured a popular economy rooted in resistance, repeasantization and agroecology. Interventions by changing regimes have had a range of effects on the growth of agroecology—often unintended.

In a classic example of neo-colonial development, Nicaragua was invaded and occupied by the United States several times, while the coffee, tobacco, banana and cotton booms that defined its early and mid-twentieth century agriculture were built around US capital during five decades of rule by the Somoza family. The rural oligarchy enforced violent sharecropping relationships through armed 'white guards' and the National Guard systematically eliminated social and community leaders. A popular insurrection toppled the Somoza regime, and the Sandinista Revolution (1979–1990) carried out an extensive agrarian reform, targeting *latifundio* landlord estates and redistributing more than half the nation's farmland. Across the country, migrant farm workers became cooperative landholders. However, the revolutionary government also promoted a chemical-

(continued)

Box 11.1   (continued)

intensive export agriculture model, in order to sell commodities to Eastern bloc countries and finance the war against the CIA-funded Contra armies that carried out terror attacks in the countryside.

It was in this context that the peasant-to-peasant method (see Box 5.2 in Chap. 5) to spread agroecological food production practices became popular in Nicaragua. Mexican agroecologist Jaime Morales, who served as an internationalist in Nicaragua in the 1980s, recounts why: "if the government sent fertilizer or pesticides to the rural areas, the shipment would be targeted by the Contras. If they sent a technician, the person would be shot by the Contras. *Campesino-a-campesino* was the only way to improve food production in war conditions."

A war-weary population voted the Sandinista Front out of power in 1990. The incoming neoliberal regime privatized education and health care and reversed land reform. The mass of rural workers who had gained land access carried out a process of reconciliation with small farmers, retired soldiers and the former Contra fighters in order to found a national chapter of the peasant movement La Vía Campesina. Workers and peasants carried out highly precarious land occupations to prevent the privatization of state lands.

In the context of structural adjustment and privatization that characterized the 1990s, the role of large non-governmental organizations (NGOs) and development agencies became fundamental in the national economy, as Nicaragua became the second-poorest country in the Western Hemisphere. At the height of the neoliberal period, illiteracy, hunger and destitution reigned in Nicaraguan cities and in the countryside. Meanwhile, the campesino a campesino agroecological movement grew exponentially in the context of resistance to neoliberal reforms in the countryside and fuelled by the huge budgets of foreign development agencies such as the Ford Foundation (McCune and Sánchez 2018).

By the early 2000s, Latin America was beginning to experience the renewed influence of socialism, and the governments of Cuba and Venezuela supported municipal politicians in Nicaragua to carry

(*continued*)

Box 11.1   (continued)

out pilot programmes to support health, education and peasant agriculture. In 2006, Sandinista leader Daniel Ortega won the presidential election. Institutional support for agroecology was formalized when the National Assembly passed Law 693—The Food and Nutritional Security and Sovereignty Act—in 2009 and Law 765—The Foment of Agroecological and Organic Production Act—in 2011. However, many observers have noted how in the drafting process, disagreements between international NGOs and local partners, the Food and Agriculture Organization of the United Nations (FAO), political parties and national agribusiness, led to the watering down of the language (Godek 2015) and the lack of budgetary commitment to implementation (Schiller et al. 2019). What has been noted with less frequency is that the overall political economy being created in Nicaragua since 2007 has led to a net repeasantization, with the rural population growing faster than the urban population.

The large part of the population that came to cooperatively own means of production (including land, vehicles, machinery, licences to operate, and credit), either through the redistributions of the 1980s or the worker- and peasant-led occupations of the 1990s, has consolidated what in Nicaragua is known as the popular economy: a non-state, non-private agricultural sector that is based on associative and self-managed economic activity. This sector uses the means of production to produce employment for itself, often through agroecological practices and by de-commodifying labour relationships (Núñez 2000).

"Peace in our places of origin" was the key, according to Edgardo Garcia, General Secretary of the Land Workers Association (ATC-Nicaragua) and currently a member of the International Coordinating Committee of La Vía Campesina. "We shifted from a state of siege in our farms and communities to a situation of stability. We won land now titles, titles for our cooperatives, access to local markets." The state now provides low-interest loans to peasant cooperatives, the National Institute of Agricultural Technology has reoriented its work toward promoting native and creole seeds and agroecological

(continued)

Box 11.1 (continued)

practices, and adult education based on skill-sharing to tens of thousands of students is taking place in nearly one thousand rural communities (Osejo 2014).

Perhaps most importantly, national institutions cooperate at the local level through municipal and departmental Production, Commercialization and Consumption Councils, sharing vehicles and personnel with one another, universities and peasant organizations in order to meet the local demand for workshops, schools and social movement gatherings, even providing locally significant services such as emergency transportation in the countryside and mobilizing around climate disasters. Women's equality has also emerged as a major component of the Sandinista regime in Nicaragua, with the country holding steady in the fifth place for gender equality in the world since 2016 (World Economic Forum (WEF) 2020).

In 2018, an extremely violent political conflict broke out in Nicaragua when labour unions and the government decided to raise a social security tax on businesses. The country polarized for months as scores of people were killed in disputed circumstances.

García explains how in the midst of this crisis, the importance of peasant agroecological farming became evident: "In 2018, when the large companies called for an economic shutdown, the peasant markets across the country remained open. When they wanted people to see that there was no way to eat, people instead saw us. And they saw that the large capitalist agro-food complex had become a parasite." A regime of economic sanctions was imposed upon Nicaragua in 2018 by the United States and the European Union and the country has been denied access to credit, even in the context of the COVID public health crisis.

In this more recent situation of reduced foreign investment and private sector employment, the agroecological transformation has accelerated. Diversified, small farms have been emphasized in the anti-imperialist discourse of the national government and supported by state policies. While genetically modified seeds continue to be prohibited in Nicaragua, around 400 community seed banks provide native

(*continued*)

Box 11.1   (continued)

and open-pollination seeds to farmers through barter and seed loan systems (McCune 2016). The country has gone from producing 30% of the rice it consumes to producing over 80% in just over a decade, and it is self-sufficient in beans, the other main staple, as well as fruits, vegetables, meat and dairy products, with 80% of food produced by smallholders (Centeno 2020). Edgardo García summarizes creativity that led to the growth of peasant-led agroecology in Nicaragua as follows: "when there is a force that wants to paralyze you, block you, and leave you dead, you have to look for ways forward."

This box was prepared by Nils McCune.

Continuous reflection is therefore critical in order to assess on an ongoing basis whether interventions are still supportive of a transformative agroecology or additional steering or a new intervention is required. This is what we refer to as reflexive governance. In order to embed reflexive governance within the transformation process, *ongoing and iterative social learning processes* among actors is needed. In this way, actors can enable and ensure governance interventions that both nurture and anchor agroecology.

Agroecological transformation therefore demands governance and facilitation mechanisms that enable continuous discussions, negotiations, exchange and joint planning between actors. This will also contribute to maintaining momentum and longevity (see Box 11.2). Facilitators in this case act not as expert intermediaries but rather as enablers of local and trans-local processes. It is their task to continuously reflect on the effects of any intervention and to ensure that agricultural producers and citizens in their respective territories control transitions. In that way, efforts towards institutionalization can be kept in check by the grounded realities, possibilities, needs and agency of these people, in a democratic and socially just process. Such participatory governance must ensure that time, resources, expertise and coalition-building are organized in a way that minimizes existing power imbalances (Peuch and Osinski 2019).

There are several challenges in participatory governance processes, as highlighted by Koen Kusters et al. (2017). Because these processes involve complex relations between multiple stakeholders and their heterogeneous interests, they require significant time commitment from participants to resolve differences as well as resources (including financial) to mobilize actors and maintain commitment. Moreover, the most well-positioned institutional host or facilitators of these processes may not be the producers, or producers' organizations, involved in agroecology but rather researchers or experts in organizational development. They often have access to resources, salaries, connections and competencies, and trust from regime actors, to carry out this work. However, for many historic reasons, agricultural producers, as the key protagonists of agroecology, are highly cautious of the power of external experts and professionals to drive and control processes in ways that may not always reflect their priorities.

Indeed, as we have seen elsewhere, scientists, policy-makers, NGOs, consultants, institutions and funding processes often effectively reproduce the power relations and dynamics from the dominant regime that undermine the perspective, agency and voice of food producers. Thus, participatory governance in agroecology must be 'endogenous', or steered *from within*, instead of driven by external actors or objectives. This involves reversing the current democratic deficit which excludes

---

Box 11.2 Five Steps for the Reflexive and Participatory Governance of Agroecological Transformation

Michel Duru et al. (2015), among others, offer a framework for a 'co-innovation' process for managing and governing agroecology transformations. They outline five steps, adapted below as key questions, that can be used to effectively facilitate processes for coordinating actors at the local or territorial level.

(1) Who are we?

The question of who is, or should be, involved in the agroecology transformation in the territory is key. Important players in this context are food producers from different backgrounds, genders, castes, sectors and classes, as well as supporting actors in research, government, civil society, the private sector and media.

(2) Where are we in the process?

(continued)

Box 11.2 (continued)

Participants analyse the current situation in the territory, identifying the key assets for agroecology in the territory along with the barriers that limit it. This process often involves using participatory approaches to map out the available social and material resources and other elements of the territorial capital and/or joint development of the history of the territory. This stage may involve identifying exogenous changes and drivers that may influence the local situation, to build a collective understanding of the state of play. The exercise could provide a baseline for ongoing evaluation.

(3) Where do we want to go?

Using forecasting, participants design a future territorial organization to support agroecology. This generally involves a process of collective vision development and negotiation of values, aspirations and interests.

(4) How do we get there?

Using backcasting, participants identify the steps required to move from the current context to the vision identified in step 3, building on what has worked in the past. It includes developing an understanding of which additional assets, resources, contacts, competencies and capacities are required to achieve the desired change while recognizing that the 'desired change' may be contested and may evolve over time.

(5) What have we learned?

In this stage, participants set up decision-making processes and strategies for ongoing participatory governance of the transformation process, including systematically and iteratively monitoring progress against the objectives. From here, the process could circle back to point 1.

Source: Duru et al. (2015)

food producers and citizens and favours the interests of powerful actors such as corporations and technocratic government. It often requires an expansion of 'direct' democracy in decision-making in order to complement, or replace, models of representative democracy that prevail in conventional policy-making.

This is a major challenge. First, deepening democracy assumes that every citizen is competent and reasonable enough to participate in political decision-making. However, for some people this requires a life-long education process to develop a different kind of character from that of passive taxpayers and voters. Second, empowering food producers as well as other citizens in the governance of food systems requires social innovations that: (1) create safe spaces for deliberation and inclusion and that contribute to gender equity, as well as other forms of equity; (2) build local organizations, horizontal networks and federations to enhance peoples' capacity for voice and agency; (3) strengthen civil society; (4) expand information democracy and citizen-controlled media (e.g. community radio and participatory film-making); (5) promote self-management structures at the workplace and democracy in households; (6) learn from the rich history of direct democracy; and (7) nurture active citizenship (Pimbert 2012b). Last, an endogenous steering of this kind of democratic governance for agroecology is driven by the idea that active participation in decision-making is a right that is claimed through the agency of people themselves; it is not granted by the state or the market.

The Territorial Governance of Agroecology Transformations

While agroecological governance spans scales from household and farm to national and international, the territory is increasingly seen as the decisive level in fostering agroecological transformations (e.g. Wezel et al. 2015). Territories are not (only) delineated by administrative boundaries. Rather, they are generally defined by a range of circumstances and context- specific factors: spatial, geo-physical and environmental conditions, political and administrative structures, and cultural identities. Key aspects of a territorial approach include the valorization of endogenous resources, intersectoral development, the recognition and celebration of local identities, self-control of "development processes" and solidarity and democracy (Wezel et al. 2015).

It is evident then that the territory (or landscape level) is important for agroecology. It is at this scale that direct interactions take place between ecological and social processes that (re)connect agriculture, food, environment and health (Lamine et al. 2019). A territorial approach to agroecology thus allows for holistic perspectives that take into account interlinkages between the three dimensions of sustainable development—social, economic and environmental—and the possible tensions and trade-offs between these dimensions and between different sectors. In other words: in the territory, farm-level land-use decisions that involve ecosystem functions (i.e. pollination and watershed management) are connected with dynamics at a landscape or territorial level (Wilson 2009). Key to the potential for agroecological transformation is thus interaction and collaboration between food producers and other land users in a territory.

The territorial scale, like that of the local community, is intimate and rooted in place, enabling people to build "a collective attachment to a community of fate" (Lamine et al. 2019, p. 13). At the same time, it is large enough to allow for more robust mobilization of collective resources. Éric Sabourin et al. (2018) underlined this same thinking as follows:

> The proposals of support for the development of agroecological agriculture need to be formulated at the scale of the territory and not of the technical system of the production unit or even less at the scale of the cultivated plot. The territory is the scale of the management of natural resources and landscapes, social life, knowledge management networks and local, regional and national markets.

A focus on the territory also provides opportunities to shift from linear and globalized commodity-supply chains towards locally controlled circular systems reintegrating food and energy production with water and waste management. This can be achieved by closing nutrient cycles, using functional biodiversity and ensuring that production, distribution and consumption are established within the territory (Pimbert 2012a). Circular systems significantly reduce fossil-fuel use and emissions; boost food, water and energy security; create jobs; raise incomes; promote resilient and self-reliant communities; and enhance the potential for inclusive and democratic governance (Jones et al. 2012). The regeneration of agroecology and circular systems within territories thus contributes to an improved 'quality of life' and helps in meeting the Sustainable Development Goals.

In terms of power and governance, the territory is an important interface between top-down provisioning by government programmes and investment *and* the democratic expression of citizens' needs, aspirations and demands—it is precisely here that the two can mesh through issues of power and governance (van der Ploeg 2018). At the territorial scale, support structures and resources can be tailored to specificities of place (OECD/FAO/UNCDF 2016) while increasing the potential for building and mobilizing territorial resources and mechanisms (knowledge, labour, relations, nature) to further catalyse agroecological transformations. Thus, the territory allows collective work to shift the rules of the game, reform institutions, build markets and foster innovation. Not surprisingly, some of the most successful examples of agroecological transformations in this book are the result of such a territorial approach.

The territorial governance approach to agroecology can be strengthened in some ways through regional institutions and through regional policy (FAO 2018; Wezel et al. 2015; Petersen 2017) but also through new grassroots and alternative institutions that transcend existing regional boundaries. For example, as we have argued in the chapter on systems of economic exchange (Chap. 6), part of building an agroecological system involves developing territorial and interterritorial markets, distribution mechanisms and processing facilities—from mills and local abattoirs to community-owned food-processing units—because foodstuffs produced via agroecological methods are often ill-suited for undifferentiated export markets. Experiments with new institutional arrangements, such as food policy councils or the biodistricts in Italy, are exemplary developments where new territorially based institutions are carving out new strategic roles, convening multiple stakeholders in a territory (see Box 11.3). Such new institutions and grassroots networks help to broaden and deepen (Petersen 2017) territorial connections, relations and practices within the multi-scale governance framework (Fig. 11.2) and are most effectively built through the agency of territorial actors in processes of endogenous development.

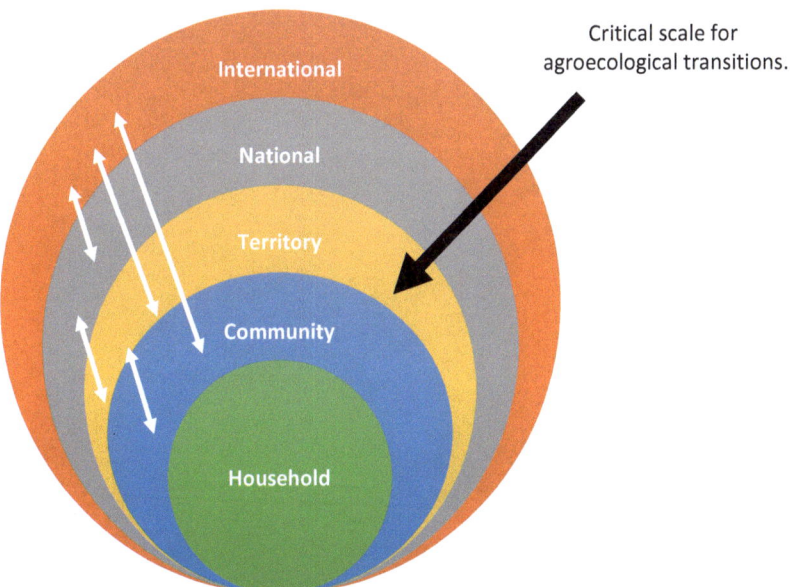

Fig. 11.2 Agroecology should be considered within a multi-scalar governance framework that examines the dynamic relationship between actors, institutions, systems and policies across household, community, territorial, national and international scales. At the same time, there is growing evidence over the importance of the territorial scale for agroecological transitions

Box 11.3   Biodistricts in Italy—Agroecology Transformations at the Territorial Scale

In Italy, the 'biodistrict' model was first launched by a farmers' organization, the Association for Organic Agriculture, in 2009. Biodistricts convene multiple stakeholders in a territorial space to advance the local management of natural resources based on the principles of organic agriculture. These initiatives focus on governance, aiming to strengthen the interlinkages between actors such as farmers, consumers, the touristic sector, municipalities, regional parks and other local associations to improve local economic, social and ecological conditions. There are now dozens of biodistricts in Italy, each emphasizing and valuing place-based cultures and mobi-

(continued)

Box 11.3   (continued)

Fig. 11.3   Farmers from around the world tour the biodistrict della Via Amerina e delle Forre as part of the Schola Campesina international learning exchange in Italy (*Photo Credit:* Colin Anderson)

lizing territorial capital to provide new employment and livelihood opportunities, improve ecological conditions, attract people to rural areas and foster the production of territorial and often traditional products.

Importantly, these have a strong basis in farmers' organizations but have gained support from municipal governments and other actors in a promising territorial approach. The European Network and Mediterranean Biodistricts are playing an important role in sharing this innovative model with others, particularly in Europe, to help with interterritorial sharing and collaboration (Fig. 11.3).

Source: International Network of Eco-Regions (2017)

REFERENCES

Centeno, E. (2020, March 31). Entrevista con el Ministro de Agricultura. *Revista En Vivo.*

Duru, M., Therond, O., & Fares, M. h. (2015). Designing Agroecological Transitions; A Review. *Agronomy for Sustainable Development, 35*(4), 1237–1257.

FAO. (2018). *Catalysing Dialogue and Cooperation to Scale Up Agroecology: Outcomes of the FAO Regional Seminars on Agroecology.* Rome: FAO.

Godek, W. (2015). Challenges for Food Sovereignty Policy Making: The Case of Nicaragua's Law 693. *Third World Quarterly, 36*(3), 526–543.

International Network of Eco-Regions. (2017). *52 Profiles on Agroecology: The Experience of Bio-Districts in Italy.* FAO.

Jones, A. D., Pimbert, M. P., & Jiggins, J. (2012). *Virtuous Circles: Values, Systems, Sustainability.* London and Geneva: IIED and IUCN.

Kusters, K., Buck, L., de Graaf, M., Minang, P., van Oosten, C., & Zagt, R. (2017). Participatory Planning, Monitoring and Evaluation of Multi-Stakeholder Platforms in Integrated Landscape Initiatives. *Environmental Management, 62*(1), 170–181.

Lamine, C., Magda, D., & Amiot, M.-J. (2019). Crossing Sociological, Ecological, and Nutritional Perspectives on Agrifood Systems Transitions: Towards a Transdisciplinary Territorial Approach. *Sustainability, 11*(5), 1284.

McCune, N. (2016). Family, Territory, Nation: Post-Neoliberal Agroecological Scaling in Nicaragua. *Food Chain, 6*(2), 92–106.

McCune, N., & Sánchez, M. (2018). Teaching the Territory: Agroecological Pedagogy and Popular Movements. *Agriculture and Human Values., 36*(3), 595–610.

Núñez, O. (2000). *La economía popular en Nicaragua.* Managua: CIPRES.

OECD/FAO/UNCDF. (2016). *Adopting a Territorial Approach to Food Security and Nutrition Policy.* Paris.

Osejo, N. (2014). *Escuelas de campo para transformar la realidad productiva de Nicaragua.* Lima: National Agrarian University.

Petersen, P. F. (2017). *Arreglos institucionales para la intensificación agroecológica. Una mirada al caso Brasileño desde la agroecología política.* Tesis doctoral, Universidad de Pablo de Olavide, Sevilla.

Peuch, J., & Osinski, A. (2019). *Governing the Transformation of regional Food Systems: the Case of the Walloon Participatory Process.* Working Paper: UCLouvain Institute for Interdisciplinary Research in Legal sciences.

Pimbert, M. P. (2012a). *Fair and Sustainable Food Systems: From Vicious Cycles to Virtuous Circles.* London: IIED.

Pimbert, M. P. (2012b). *Putting Citizens at the Heart of Food System Governance.* IIED Policy Briefing. London: London.

Sabourin, E., Le Coq, J. F., Fréguin-Gresh, S., Marzin, J., Bonin, M., Patrouilleau, M. M., Vázquez, L., & Niederle, P. (2018). *Public Policies to Support Agroecology in Latin America and the Caribbean*. Montpellier: CIRAD.

Schiller, K., Godek, W., Klerkx, L., & Poortvliet, P. M. (2019). Nicaragua's Agroecological Transition: Transformation or Reconfiguration of the Agri-Food Regime? *Agroecology and Sustainable Food Systems, 44*(5), 1–18.

van der Ploeg, J. D. (2018). *The New Peasantries: Rural Development in Times of Globalization*. London and Sterling, VA: Earthscan Food and Agriculture.

Wezel, A., Brives, H., Casagrande, M., Clément, C., Dufour, A., & Vandenbroucke, P. (2015). Agroecology Territories: Places for Sustainable Agricultural and Food Systems and Biodiversity Conservation. *Agroecology and Sustainable Food Systems, 40*(2), 132–144.

Wilson, G. A. (2009). The Spatiality of Multifunctional Agriculture: A Human Geography Perspective. *Geoforum, 40*(2), 269–280.

World Economic Forum (WEF). (2020). *Global Gender Gap Report 2020*.

CHAPTER 12

Conclusion

Abstract The final chapter concludes the book by summarizing our arguments and the urgency of agroecology transformations.

As the world's crises exacerbate inequity and fuel the erosion of the ecological basis of the world, the urgent need for transformative change is palpable. Agroecology responds to this call for change. Our formulation of agroecological transformation reflects not one grand theory of change but a recognition of a co-evolutionary and adaptive approach. It also underpins the importance of collective action, social movements and solidarity networks as a means of building and amplifying political power and community agency to advance agroecology transformations.

Keywords Agroecology • Crisis • Future

As we are bombarded with news of multiple intersecting food system-related crises—hunger, pandemic, climate change, biodiversity collapse and gross inequity—the edifice of the corporate industrial global food system is crumbling. Peasants, indigenous peoples, women, black and people of colour, among other groups and peoples, have long lived at the sharp end of the colonial-corporate stick. Now, as these crises exacerbate inequity and fuel the erosion of the ecological basis of the world, the urgent need for transformative change is palpable and calls for change grow louder.

© The Author(s) 2021   191
C. R. Anderson et al., *Agroecology Now!*,
https://doi.org/10.1007/978-3-030-61315-0_12

Agroecology responds to many of these crises and offers multiple benefits (see Chap. 2): enhancing biodiversity, addressing climate change, contributing to good nutrition, strengthening social relations and—in its most radical and most needed form—directly challenging coloniality, inequity and oppressions. Social movements have been advancing agroecology as a paradigm for food systems that centres the voice, agency and priorities of these often-marginalized peoples. We have seen how—far from merely a tweaking of the existing system—political agroecology is rooted in the politics of food sovereignty. It simultaneously rejects the dominant food regime while offering an alternative vision and a pragmatic and viable set of principles as the basis for transformation.

The urgent need to advance this paradigm is why we chose the title for this book: *Agroecology Now!* We have sought to articulate what the processes of agroecology transformation look like at this historical juncture. Agroecology is an idea whose time has come. The need for transformation is laid bare, the idea has been foregrounded by social movements, local and territorial experiences in advancing the system are coalescing, and adjacent social movements from Black Lives Matter to climate justice and the World March of Women are gaining momentum.

At the same time, gross inequity has deepened, as has the vested power of the elite. While agroecology provides a promising alternative paradigm for food systems, there are tremendous barriers that prevent the transformation, which we have outlined in depth throughout the book. The disproportionate power wielded by the architects and beneficiaries of the dominant regime over food governance underlies most of the lock-ins and barriers to agroecology.

In Part I, we defined agroecology as a process of continuous transition based on core principles and a political commitment to both social justice and ecological regeneration. Contrary to what is sometimes thought, agroecology is not just a set of technical practices, but its transformative potential is grounded in its social, cultural and political dimensions. In fact, these dimensions are what distinguish agroecology from the many competing 'solutions' that are being proposed in the form of climate-smart agriculture, the Fourth Industrial Revolution and sustainable intensification. In contrast to agroecology, these centre corporate-led approaches for short-term and marginal gains in sustainability and leave in place the profit-centred logics and structural inequality that prevent the flourishing of nature and humanity.

At its root, agroecology is based on a shift in political and economic power from corporations, governments and elites to food producers and other citizens. It emphasizes production and distribution processes that are self-reliant and thus have limited commercial and speculative value for financial institutions and the shareholders of agri-food corporations. It enshrines the collective knowledge of food producers—especially women—and thus requires a fundamental change from dominant Western and patriarchal expert-driven knowledge and development systems. People's knowledge and agency are central to agroecology and is prioritized and brought into dialogue with scientific knowledge and other ways of knowing in a political agroecology approach.

Learning from the growing number of local experiences, case studies and critical analyses of agroecology in different parts of the world, we explored agroecology transformations as emergent, non-linear, context-specific and messy processes. We adopted the Multi-Level Perspective on Sustainability Transition to conceptualize agroecology not only as a niche but also as a proto- and counter-paradigm for food systems that is being advanced through political action at multiple scales. Our analysis explained how agroecology transformations occur at the various points of intersection and contestation between agroecology (as the 'niche' level) and the 'regime' in the six domains of transformation: (a) Access and Rights to Nature; (b) Knowledge and Culture; (c) Systems of Economic Exchange; (d) Networks; (e) Equity and (f) Discourse. In each of the chapters in Part II we unearthed the factors and dynamics that limit agroecology transformations and drew out examples and dynamics where agroecology transformations are enabled.

While many studies have drawn out some of the enabling factors or drivers in one or more of these domains, or emphasized the disabling ones, we looked across these studies to impute patterns that emerged across the experiences of agroecology transformations in different settings. This provided the basis to systematically and simultaneously articulate the enabling and disabling conditions within each of the six domains of transformation that emerged. While we identified six discrete domains in agroecology transformations, we emphasize that transformations will not be possible through a reductionist approach. It is essential that intentional processes of agroecological transformations not reduce action to singular domains—such as creating new markets (a common refrain)—but to consider and support transformations at the intersection of these multiple domains.

In Part III of the book, we drill down on the notion of governance interventions and what effects different approaches can have on agroecology transformations. We observed six effects in and across domains of transformation. This six-part framework provides nuance to the often-binary division between interventions that encourage conforming to the dominant regime and those that transform it. Our six effects of governance interventions are placed on a spectrum from those that directly supress agroecology to those that dismantle the regime and strengthen agroecology.

We raised two complementary effects of governance interventions that undergird agroecology transformations. One is *nurturing* of agroecology—which includes interventions that support self-managed, grassroots networks and communities to develop agroecology on their own terms, rather than to conform to broader economic or political agendas that derive from the logics of the dominant regime (as is often the case with government policy and programmes). These often are the result of participatory and inclusive processes for policy-making and institutional choices, organizing citizens for widespread democratic coordination and beyond. The other promising governance effect is *releasing* agroecology from the disabling conditions of the dominant regime through contesting norms, structures and practices that disable agroecology while simultaneously anchoring it in the regime.

The outcomes or effects of any type of governance intervention (e.g. a policy, programme, project) are however not static or constant. For example, interventions that are intended to support agroecology may end up co-opting or containing it. To this end, in Part III, we argue that a participatory and continuously reflective approach is vital. We also articulate the territory as a critical yet underappreciated scale at which agroecology transformations can be supported.

Governments have in some cases played an important role in agroecology and especially have an important role in limiting the power of dominant regime actors. Yet, agroecology follows a bottom-up logic that is diametrically opposed to the systems of elite governance in place in many or most countries. Political agroecology is congruent with deeper forms of democracy which include civil society participation in decision-making, participatory democracy, and community self-organization in territories. Agroecology transformations thus fundamentally challenge governments and wider society to adopt forms of governance that counter current uniformity, centralization, blueprint planning, control and coercion.

Our formulation of agroecological transformation reflects not one grand theory of change but a recognition of a co-evolutionary and adaptive approach that involves multiple transformations. It also underpins the importance of collective action, social movements and solidarity networks as a means of building and amplifying political power and community agency to advance agroecology transformations. This is easier said than done. But, given the threats—from climate change and disempowering political dynamics to challenges to food security—it is arguably the most viable and socially just pathway to food systems fit for the challenges and opportunities of our tumultuous times.

INDEX

© The Author(s) 2021
C. R. Anderson et al., *Agroecology Now!*,
https://doi.org/10.1007/978-3-030-61315-0